The Earth Is Our Home

Mary Midgley's Critique and Reconstruction of Evolution and Its Meanings

Nelson Rivera

www.imprint-academic.com

Copyright © Nelson Rivera, 2010

The moral rights of the author have been asserted
No part of any contribution may be reproduced in any form
without permission, except for the quotation of brief passages
in criticism and discussion.

Published in the UK by Imprint Academic
PO Box 200, Exeter EX5 5YX, UK

Published in the USA by Imprint Academic
Philosophy Documentation Center
PO Box 7147, Charlottesville, VA 22906-7147, USA

ISBN 978 1845402129

A CIP catalogue record for this book is available from the
British Library and US Library of Congress

To Sara,
and to our daughters
Noelia, Paula, Celeste, and Laura,
in love and gratitude

Contents

Introduction 1
1. Midgley in Context 9
2. What Concerns Midgley 43
3. How Come Evolution? 89
4. Travails of Evolutionary Theory with Religion . . 129
5. Midgley's Approach: Thinking from Below 171
   Bibliography 217
   Index 235

Introduction

This book attempts to show how an evolutionary theory is truly a welcome development in philosophical and religious thinking alike. This I do by assessing the contributions of philosopher of science and culture Mary Midgley (b. 1919) to an epistemology that takes Darwinian evolution seriously by building knowledge about the world from the complexity of the human experience. What Midgley has to offer is a kind of "evolutionary epistemology." In her case, as she herself puts it, that means "asking what we know by asking about ourselves as knowers, and as beings who are shaped by a particular evolutionary history, rather than by trying immediately to understand the subject-matter that we know (or don't know) about."[1]

It is my personal conviction that Midgley's approach can also be understood as a way of building knowledge of the world around us and of ourselves "from below," from the underside of the world, from the realms of nature and history.[2] We learn about ourselves in relation to everything else, especially other living things. Moreover, we learn

---

[1] Personal communication between Mary Midgley and myself dated on February 28, 2006.

[2] The image of building knowledge "from below" is my reading of Midgley's approach to evolutionary epistemology. The reference "from below" is rather commonplace is theological studies, first in Christological studies but also in Liberation theology. I explain more about it in the context of chapter 5.

about the world by using historical thinking. We build our worldviews by the gradual accumulation of experience, reasoning, and the allocation of value borne by our inner life. In this project, I also want to explore the impact that such a view "from below" may have on helping the public in general, and religious believers in particular, come to terms with evolutionary theory.

It is my belief that some of the difficulties that a number of Christians may have with evolutionary theory spring from a religious epistemology that begins with assumptions about God and the world "from above," from a previous metaphysical commitment to ideas of order, design and purpose in nature, which may impede a fuller appreciation of what a Darwinian evolutionary perspective has to offer. All that said, the view from below, which is a historical perspective, is not necessarily alien to metaphysical constructions. When a historical approach goes beyond the original experience, it then ends up being metaphysical at least in a general sense. But starting with a metaphysical theory of reality is not the most appropriate way to build knowledge of the world, at least methodologically speaking. We construct a reasonable worldview first by taking seriously our concrete experience of the natural world, including elaborating a workable picture of its evolutionary history.

In the context of this project, I first need to introduce Mary Midgley and why I think that she matters for our study of evolution and its meanings. Unfortunately, Midgley is not as well known in America, where I live and work, as she is in her native England. Therefore, I use the first chapter to present her to the general public by opening a view to her context: family, education, formative experiences, and some of the interests that have driven her academic as well as her public life.

In the same line, the second chapter summarizes those topics and concerns that have preoccupied Midgley's career. Of particular interest for our exposition is to make clear that Midgley's concerns are about science and its role in contemporary societies. Throughout this book, I intend to explain what it is that Midgley understands by science, and what problems she sees with a certain kind of science (some would call it 'scientism') that tends to overreach by intending to explain more than it actually can, or to offer more than it can actually deliver. In order to do this, I shall concentrate primarily, though not exclusively, on those of Midgley's writings dealing directly with science, evolution, Darwin, religion, and related topics.

Chapter three is a survey of evolutionary theory with a brief account of the major questions and debates that have surrounded it, roughly, since Darwin's day. If we want to talk about evolution, we had better understand what it is and what the experts are saying about it nowadays. The end result of this third chapter is a presentation of a "pluralistic understanding of evolution" akin to Midgley's own.

The fourth chapter then introduces some of the relevant controversies between evolution and religion, with special attention to those instances in which evolutionary theory has been presented, and in some cases even defended, as a form of religion itself, or at least a form of spirituality. In this chapter, I begin to draw on Midgley's critique of evolution (as a worldview) and its possible meanings. In order to do this, I revisit first some of the controversies around Darwinian evolution during the Victorian era. Most importantly, and following Midgley's lead, I then examine some of the controversies around modern sociobiology, and some aspects of the scientific materialism of the likes of Edward Wilson and Richard Dawkins, both of them *bona fide* scientists who have written considerably on the critique, sometimes dismissal, of religion by evolutionary theory.

In the fifth and final chapter, I attempt to explain what an evolutionary epistemology looks like from Midgley's perspective. Midgley herself has acknowledged that the term "evolutionary epistemology," although not used by her in her own writings, nonetheless applies well to her views on human knowing and the human interrelation with the living world. For this, I have to assess first the claims of Midgley's empiricism, since the latter is basic to the development of her philosophical epistemology. In addition, I try to answer the question of why it matters to elaborate such views based on Midgley's work. Midgley herself has insisted on the necessity of our self-understanding as creatures who are products of this earth, who have our home here, and cannot live apart from this intimate relation with nature. That is the reason for the use of the image that appears in the title of this book: the earth as home.

For Midgley, both science and religion are needed in our way to wisdom. Neither of them can take the place of the other in their particular approaches to the world. Natural science is a way of getting acquainted with the living world in all its complexity. The world is an intricate web of relations; it is also the realm of wonder. Wisdom, therefore, comes into a way of life that can be appropriately informed by science but that goes beyond science and into ethics, art, metaphysics, and, yes, religion. Religion provides a sense of direction for life. Religion does it by giving us a fuller view of the human experience. Life is not merely the aggregate of individual parts, but rather the foundation for the appropriation of the human experience in its wholeness. Therefore religion and science are together capable of providing us a measured sense of our own selves, for instance, our innermost desire to know who we are and what is our place and role in this earthly life.

During the time of my research and writing, I was blessed with the opportunity of an exchange of letters (electronic mail) and therefore views with this living philosopher. Since her first major publication, *Beast and Man*, in 1978, Midgley has earned a reputation as one of the most effective "scourges of scientific pretensions" in our time. This is because she has criticized the ideologies that unfortunately sometimes drive scientists in their supposedly "objective" and "unbiased" work. These are ideologies that proclaim a reductive view of the world and that tend to deprive us of the world's intrinsic richness and complexity. As a moral philosopher, Midgley has called attention to the ethical implications of systems of ideas, especially the dominant ideas and ideals of a given time, a role that has been increasingly occupied by scientific discourse in our day. Midgley is deeply concerned with the impact on the public at large of overreaching scientific ideologies and their seemingly philosophical baggage.[3]

As someone who thinks of philosophy as more than just an academic discipline or a private exercise, Midgley has been an advocate of a public role for philosophy and has called for a critical reappraisal of its social responsibility. Philosophical analysis is a valuable tool in the task of unearthing dangerous patterns of thought. In this context, Midgley engages in what she calls "philosophical plumbing," a way of uncovering the role of ideas and ideologies in forming our modern worldview. For her work in explaining difficult concepts to the general public, and the open reception and success of her many publications, Midgley has been praised as a philosopher that the common person can actually understand. At the same time, she is also said to be "feared and admired" in almost equal measure.

Among Midgley's many contributions to the philosophy of science, I find particularly insightful her assessment of evolutionary ideas. She is able to be as fiercely critical of developments in evolutionary theory, and their impact on culture, as she can be affirmative of the theory's contributions to the history of ideas. Moreover, and as I intend to show through this project, she has contributed positively to the reassessment of Darwinian evolution and its meanings.

In her writings on evolution, Midgley has argued that modern science and philosophy have both suffered from a kind of blindness about the earth, its significance and pre-

---

[3] See David Midgley, ed., *The Essential Mary Midgley* (London: Routledge, 2005), 2.

dicament. The earth is a living entity, not the imperturbable object of human activity and abuse. For us, the earth is more than context and environment: it is our home. We are of this earth, that is, we belong here and, therefore, cannot think of ourselves as mere visitors or passengers in it.[4] We are an intrinsic part of nature; we are part of the whole. But in order to comprehend this truth and its implications for life and thought, we need the proper tools of analysis. We need a proper way of looking at things, a way of building knowledge, or an epistemology.

For Midgley, this is where Darwinian evolution, as a theory about nature and life's processes, becomes handy. In her view, Charles Darwin (1809–1882) has given scientists and philosophers — and I want to add theologians to the mix — a "down-to-earth" approach to building knowledge about the world as well as a sensible way of relating to that world. All that said, Midgley also warns us that evolution does not necessarily have an exclusive or unified understanding of the world. Evolution has multiple meanings and applications. Evolutionary thinking itself is rather a complex web of theories and perspectives. Midgley herself embraces a pluralistic view. "Darwinism is often seen — and indeed often presented — not as a wide-ranging set of useful suggestions about our mysterious history, but as a slick, reductive ideology."[5] In order to understand the multiple meanings of evolution, the Darwinian image of the "branching tree" is probably the most appropriate.

We could say that Darwin is one of Midgley's heroes. Midgley's respect and appreciation for Darwin's life and work are very much present throughout her writings. Like Darwin before her, Midgley's holistic thinking brings together a proper attention to the natural world, a flair for

---

[4] See James Lovelock's foreword in *The Essential Mary Midgley*, viii.

[5] Mary Midgley, *The Ethical Primate: Humans, Freedom and Morality* (London: Routledge, 1994), 17.

historical thinking, and a sense of reverence for thought and life in equal measure—not as separate things or autonomous spheres of learning but as essentially and intrinsically related parts of a whole. All these elements come together in a tangled web of connections. And this is precisely where she finds Darwinian evolutionary thinking so helpful.

This is not to say that Darwin had it all figured out, or that he was always consistent in his thought. As Midgley for one has argued, Darwin was too selective in his doubts, say, concerning religion. In Midgley's opinion, this was a real limitation, if not an open flaw, in Darwin's thought. Yet she can still celebrate one of Darwin's greatest contributions, not just to science, but to philosophy and religion as well: an attitude of reverence and wonder for nature, not in the general and abstract sense, but in the specific and ordinary. Everything matters in the natural world, from the "lower forms," that is, the very small and insignificant, to the seemingly "higher or complex forms," and everything in between.

Finally, it is my belief that Midgley's critique of evolution and its meanings is an indispensable work for any theologian who wants to take Darwin seriously but who also wants to understand the ends and means of God's activity in the world. Reading Midgley has helped my own attempt to understand some of the reasons for the problems that a number of people of faith continue to have with evolutionary theory. One of those problems seems to be related to methodological issues. It so happens that the way in which we proceed in building knowledge tends to determine much about our conclusions.

In the case of an evolutionary understanding of the world, unless we begin from the realms of nature and history with the least possible assumptions about purpose or design, we will have problems in coming to terms with what the theory has actually to offer regarding the richness, complexity and beauty of an evolving world, one which continues to be created. This is the world where God's presence and activity, as I believe it, allows for freedom and novelty and is, therefore, ultimately compatible with such an evolving reality.

Chapter One

Midgley in Context

Midgley's Background

Mary Midgley is surely one of England's best-known contemporary philosophers. She has been praised for her sharp mind as well as for her clear and non-technical writing style, which has allowed many to consider her philosophy as one that the common person can actually understand. Moreover, as a fiercely combative thinker, and as one that does not shy away from either academic or public controversy, Midgley has been called "the most frightening philosopher in the country, the one before whom it is least pleasant to appear a fool."[1] However, she is also said to be better at doing "demolition work" than positively explicating her own ideas. I believe that the public perception is partly due to her acute intelligence, the critical (some might say harsh) tone of her statements, and to an unabashed and creative use of language.

Born in 1919 to a middle class family living in London, Midgley grew up in an environment that stressed education and cultural enrichment. Midgley's father was an Anglican priest who served as military chaplain during the First World War, and later became chaplain at King's College, Cambridge University. Later in life, he became a convinced pacifist. Her mother came from a well-to-do family, among

---

[1] See the profile written on Midgley by British journalist Andrew Brown, "Mary, Mary, Quite Contrary" in *The Guardian*, Saturday, January 13, 2001.

them a grandfather who was a renowned engineer. Growing up in a vicarage, theirs was a household that cherished deep ethical and religious values as well as independent thinking. She was educated at Downe House School, which was located at first in the former home of Charles Darwin, now a museum. Later on Midgley attended Somerville College, Oxford University, where she was a fellow. She was awarded an honorary doctor of letters degree (D.Litt.) from the University of Durham in 1995. Besides being a prolific writer, she has identified teaching as her true vocation.

Midgley began the study of Greats (Classics) at Oxford in 1938, and then continued with philosophy in 1940, finishing her bachelor's degree in 1942.[2] During the Second World War years, given the fact that many young men had either volunteered or had been recruited to serve in the military, a number of women took advantage of the opportunity to study those disciplines that were not usually available to them (at least not in those increased numbers) or that were not tailored to their interests, like philosophy. They engaged the philosophical field, as Midgley herself has

---

[2] Midgley went back to Oxford in the summer of 1947 and began doctoral work on Plotinus' view of the soul, a topic so vast and unfashionable, by her own account, that she never finished her thesis. See her re-telling of this episode of her life in her autobiography, *The Owl of Minerva: A Memoir* (London: Routledge, 2005), 158. In a fairly recent article, she actually declares her luck in entering academic life before the current insistence on a Ph.D. as a qualification for university teaching. The reason is that, in her own words, "the Ph.D. training ... shows you how to deal with difficult arguments, which is necessary in dealing with hard subjects. But that close work doesn't help you to grasp the big questions that provide its context — the background issues out of which the small problems arose." See Midgley's article "Proud not to be a Doctor" in the *Guardian*, Monday October 3, 2005. The article is accessible through the Guardian's website: www.guardian.co.uk

said, not because they had to but because they wanted to.³ They were serious about philosophy.

Among Midgley's fellow students and friends were, to name a few recognizable ones, Iris Murdoch (1919–1999), who was in her same class, Elizabeth Anscombe (1919–2001), a year before, Philippa Foot (b. 1920), a year after, and Mary Warnock (b. 1924), who became a member of the British Parliament later in life. They all became famed philosophers in their own right. Upon their arrival, the dean of Somerville College told them that women were still on probation in Oxford, and therefore their behavior both in conduct and performance mattered.

These women belonged to a generation that struggled to come to terms with the totalitarianism of the likes of Hitler and Stalin, with existentialism à la Sartre, and also with what Peter Conradi calls the "slow collapse of organized religion."⁴ Their friendship and exchange of views continued throughout the years. In the case of Midgley and Murdoch, the conversation went on for decades.⁵

The intellectual atmosphere of the time was very much charged towards analytic philosophy. The influential book of A. J. Ayer (1910–1989), *Language, Truth and Logic*,⁶ was

---

[3] See her own comments in this regard in the chapter dedicated to Mary Midgley in Sian Griffiths, ed., *Beyond the Glass Ceiling: Forty Women Whose Ideas Shape the Modern World* (Manchester, UK: Manchester University, 1996), 163–168.

[4] Peter J. Conradi, *Iris: The Life of Iris Murdoch* (New York: Norton, 2002), xxv. This is the authoritative biography of Midgley's friend and conversation partner, the philosopher and novelist Iris Murdoch.

[5] Conradi, 84; the author details almost every significant encounter (with agreements or disagreements) between the two.

[6] A. J. Ayer, *Language, Truth and Logic* (original ed.; London: Gollancz, 1936); this book did more than any other to popularize the main ideas behind the philosophy of logical positivism among the English-speaking public.

obligatory reading at the time. Not even Midgley or her classmates could avoid dealing with that text and writing papers about it. However, they all found logical positivism and the philosophy of language, which were predominant then, more than just dry. They thought of it as being too detached from everyday life and concerns. These women were, on the one hand, quite interested in metaphysics and, on the other hand, also interested on practical applications to actual life issues.[7] At the time, ethical matters were not considered to be at the core of the philosophical enterprise, despite or maybe precisely owing to, earlier efforts by G. E. Moore (1873–1958)[8] to give moral philosophy a firmer philosophical ground.

The impact of Moore's work cannot be over estimated. The general consensus is that Moore introduced linguistic philosophy to England. His book, intended to be a revolution in moral reasoning, ended up treating ethics more as pure intuition rather than thought, especially since he believed moral judgments to be vitiated by what he called a "naturalistic fallacy." According to Midgley,[9] Moore actually narrowed the possibilities for moral philosophy as a field of enquiry by isolating it from other branches of philosophy. Moreover, his stance against old and traditional as opposed to modern ways of doing philosophical ethics may have pushed many to abandon such a pursuit. After *Principia Ethica*, all that was left to do was to further study

---

[7] See for example, the interview with Mary Midgley under the title "Murdoch and Morality," published in the book *What Philosophers Think*, ed. Julian Baggini and Jeremy Stangroom (New York: Barnes and Noble, 2003), 126.

[8] G. E. Moore, *Principia Ethica* (original ed.; Cambridge: Cambridge UP, 1903).

[9] See Mary Midgley, *Wisdom, Information and Wonder: What Is Knowledge For?* (reprint ed.; London: Routledge, 1995), especially 144–153.

instances of the "naturalistic fallacy," an unfortunate outcome of Moore's ethical (or ethico-logical) arguments. With the *Principia Ethica*, Moore seemed to have written two books in one: first a negative version, within the first five chapters, followed by a positive one in the sixth and last chapter. The former established the impossibility of using logical argumentation for ethical practices. The latter presented, in a twist to the main outline, the ethical ideal, a sort of contemplation of natural beauty and art, and an ensuing conception of the "good." These were things for his readers to keep in mind and work out, which some of his most prominent disciples did: for example, Clive Bell in his book *Art*.[10] Moore saw a close relation between morality and the experience of beauty: the beautiful is there to make the better possible. He brought back contemplation—in the Platonic sense—of goodness and beauty to the center of the moral scene.[11]

In any case, this state of affairs prompted Midgley and friends to find their own way, which included doing readings on their own, but also taking classes with Donald McKinnon (1913–1994). McKinnon was a theologian, who taught classics and analytic philosophy of religion at Oxford University for many years. Admired as a good teacher and trusted advisor by many of his students, he had strong interests also in metaphysics and ethics. Both Midgley and Murdoch have acknowledged their debt to him during their formative years at Oxford.

The Influence of Iris Murdoch

Iris Murdoch, in particular, is recognized by Midgley as one of her strongest influences. Their philosophies are said to

---

[10] Original edition in 1924; see discussion in Mary Midgley, *Heart and Mind: The Varieties of the Moral Experience* (revised ed.; London: Routledge, 2003), 69.

[11] Midgley, 72.

parallel each other in a number of ways.[12] She acknowledges having many a long discussion with Murdoch, for example, on the state of moral philosophy in England at the time, the nature of political commitment, and the desirability (or non desirability) of membership into the Communist Party (the Oxford group), among other debates. Where Murdoch's philosophy had the greatest impact on Midgley was on the importance of the inner spiritual life as seedbed for moral deliberation and pre-requisite for action. By assuming this position, she stood against Kant but also against the Utilitarian tradition.

Owing to her importance for understanding Midgley, a word about Iris Murdoch is in order here. Better known as a novelist, but also an influential philosopher in her own right, Murdoch's main contribution lies in the reclaiming of the self as individual and the re-conception of moral value as centered in the idea of the good. On the one hand, she was very critical of attempts to deprive the individual of any agency, and his or her submission to the will of the community, the state, or the dominant ideology of the day. On the other hand, she struggled to bring metaphysics back into moral deliberation and ethical enquiry.[13]

Murdoch believed that both analytic philosophy and existentialist ethics had isolated the agent from any normative framework. In Murdoch's opinion, both kinds of philosophy tended towards "solipsism:"[14] the individual stands alone, with nothing to sustain him or her, and is alone responsible for his or her own reality. The very choices that the individual makes become the only reality as

---

[12] Baggini, *What Philosophers Think*, 126, 128–130.

[13] The best study that I know of Murdoch's moral philosophy if that of Maria Antonaccio, *Picturing the Human: The Moral Thought of Iris Murdoch* (New York: Oxford, 2000). In the following pages, I am very much indebted, though not exclusively, to her analysis.

[14] Antonaccio, 25.

such. For Murdoch, neither philosophy concedes that the individual may be concerned with something real outside itself, something not of his or her own making, a given, beyond language or freedom respectively.

British empirical tradition has regarded moral beliefs, like any other "beliefs," as devoid of cognitive value. The point here is that any statement that is not supposedly subject to test or observation is held not factual and, therefore, not reliable. It is then a matter of opinion. In the latter category, we have everything from moral judgments to religious ideas, which seem to be generated by personal desires, that is, subjective values. In the empiricist's view, moral beliefs are not regarded as genuine facts about the world, but more like "attitudes" and, therefore, not a proper subject matter of philosophy.[15] Murdoch was convinced that analytic philosophy was an unfortunate departure from previous philosophy, from considering moral beliefs to be fundamental discoveries about the world and not mere attitudes. Moral beliefs and the concepts used to express them have been regarded in the history of philosophical ethics as facts about the world, rather than individual, relative, or passing judgments.

The conceptual separation of facts and values is another unfortunate development in modern philosophy. It makes all morality depend on human will, and not on something about the world. This particular dichotomy seems to flow from the aversion to any form of naturalism, an over concern with "naturalistic fallacies,"[16] which represents the analytic philosopher's attempt to separate description from evaluation in philosophical enquiry.

---

[15] Antonaccio, 29.

[16] As we have mentioned before, when referring to Moore's philosophical ethics, a view that was readily taken (and repeated *ad nauseam*) by many of his followers.

Murdoch was convinced that for a proper understanding of morality, a conception of a supreme aim or goal is needed. And nothing does better than the concept of the Good. Why? Because moral freedom is ultimately related to an idea of the good; it is for the good of the self to look after him or herself as well as for others—even though how we are to define what the good is remains a problem.[17]

In the last analysis, the good is conceived, not so much in relation to specific practices or with reference to religious traditions, but as "a realistic standard that is both internal to consciousness and also surpasses it."[18] The good is an ideal, a "perfected knowledge," through which we may try to apprehend the reality of other human beings. The idea of the good is used as a corrective for selfish illusions, a parameter of right vision, a motive or kind of force for moral well being, for living. It speaks of human nature.

Murdoch tried to preserve the self's integrity without necessarily isolating it from the reality of others who make claims and define the norms. For her, in any case, there is need to acknowledge the freedom of the self (against those who unsettle the notion of the self as agent, e.g. post-structuralism), as well as the conditions and contingencies of its situation in our search to define morality. In order to accomplish both attempts, an understanding of human consciousness is needed.

Consciousness, for Murdoch, is the fundamental mode of human moral being.[19] The activity of human consciousness refuses to be reduced or explained away by any accounts, whether these come out of "post-modern" philosophy, which denies substance to the self, or out of science, which reduces it to a deterministic outcome of causes and effects in

---

[17] See Murdoch's three seminal essays on this topic collected in *The Sovereignty of Good* (reprint ed.; London: Routledge, 2003).

[18] Antonaccio, 12.

[19] Antonaccio, 3.

its search to objectify it. Seeing a crisis in our understanding of human personality, Murdoch was anticipating much of the current agenda in ethical enquiry as, for example, in virtue theory and in moral realism.[20]

According to Murdoch, moral philosophy is properly speaking a form of metaphysics. Ethics is not necessarily separate from an understanding of reality. Concerns about the world lead to moral concerns; likewise, moral concerns lead to questions about the kind of world in which we live. The demise of metaphysical thinking in general, and of ethical thinking in particular, is the loss of our ability to theorize adequately about human reality. In the case of religion, and religious descriptions of reality, Murdoch states,

> I think that the ordinary man [sic], with the simple religious conceptions which make sense to him, has usually held a more just view of the matter than the voluntaristic philosopher... Religion normally emphasizes states of mind as well as actions, and regards states of mind as the genetic background of action: pureness of heart, meekness of spirit. Religion provides devices for the purification of states of mind.[21]

She does not deny the historical nature of moral claims and concepts. But morality also needs some kind of normative frame, a definition of reality to which its claims hold true, a body of knowledge that can verify the facts about the human situation and experience.

Murdoch's concern was then to give both a social and a particular account of the self within a metaphysical framework. Communal consensus is not enough for morality.

---

[20] Antonaccio, 4.
[21] Murdoch, *The Sovereignty of Good*, 81. This is considered by many to be her most important philosophical work. For more on these issues, see also her book *Metaphysics as a Guide to Morals* (London: Penguin Books, 1993), and the collection of her philosophical and literary writings in *Existentialists and Mystics* (London: Penguin Books, 1998).

Virtue does not stand alone, or in a vacuum, as if it were. As Murdoch puts it, "of course virtue is good habit and dutiful action. But the background condition of such habit and such action, in human beings, is a just mode of vision and a good quality of consciousness. It is a *task* to come to see the world as it is."[22] What we use is a sort of metaphysical realism, which serves to judge the validity or adequacy of historical and communal conceptions.

As I mentioned before, Mary Midgley considers Murdoch's assertion that the inner life matters as one of her major contributions. What Murdoch did was to stress the importance of an analysis of the inner life, including the full range of human cognitive capacities — which should not exclude feeling, since it is not incidental to coping with and knowing the world — in moral reflection. We need to go beyond the agent's public actions to the decisions and choices made in the struggle to become morally better, or the failure to attain moral integrity.[23] In Midgley's work, for example, motives are important in shaping not only present action but character and, therefore, future actions.

Midgley's Career

In 1950, Mary (née Scrutton) married philosophy teacher Geoffrey Midgley, with whom she has three sons — who, she proclaims, have taught her that "the human infant is not a blank paper." When Geoffrey got a job at the University of Newcastle upon Tyne, they moved there; Mary has lived there to this day. At first, she took care of her sons while editing children's books and school texts from home. In 1960, she was offered a position (originally part time) at the same university, as a lecturer in philosophy, a position she

---

[22] Murdoch, 89; emphasis is her own.
[23] Antonaccio, 14.

held for the next twenty years until her early retirement in 1980.

Mary Midgley does not consider having and raising children to be either incidental or inconsequential to philosophy. She argues that if there is something that traditional philosophy has left out, and even the Enlightenment for all its talk about the freedom and dignity of all humankind is no exception, it is the family. The non-celibate philosopher, and here Midgley mentions the names of Aristotle, Hegel, Mill, and Berkeley, is the exception to the norm.[24] She credits her sons and husband with making her reflections about human nature, human relations, and ethics deeper. Real life issues and challenges matter to philosophy, not just to ethics. In this way, conceptual and practical issues are kept together, not separate.

Something impressive about Midgley's career is the fact that she published her first book, *Beast and Man*,[25] in 1978, shortly before her sixtieth birthday, after spending a year at Cornell University in the USA, where she had been invited to teach after the publication of her well-read essay on "The Concept of Beastliness" — the book can be said to have grown out of her essay and her Cornell experience. *Beast and Man* had a fairly good reception, partly owing to the then recent controversies surrounding the publication of Edward O. Wilson's book *Sociobiology*[26] (in 1975) and Richard Dawkins' *The Selfish Gene*[27] (1976), both of which caused a great stir among academicians and which also spilled into

---

[24] Griffiths, *Beyond the Glass Ceiling*, 163.

[25] Mary Midgley, *Beast and Man: The Roots of Human Nature* (revised ed.; London: Routledge, 1995).

[26] Edward O. Wilson, *Sociobiology: The New Synthesis* (Cambridge, Mass.: Belknap/Harvard UP, 1975).

[27] Richard Dawkins, *The Selfish Gene* (revised ed.; Oxford: Oxford UP, 1999).

the general public. The stories around those two books are by now the stuff of legend.

Midgley's first book, diverse in topics and dense in arguments, was already a kind of compendium of the ideas that Midgley was going to keep elaborating and working on for years to come. After *Beast and Man*, she has published fourteen other books, her memoirs[28] and an edited book on the meaning of Gaia theory[29] being the latest additions, as well as a good number of essays and articles in books and journals. Many of those extend her arguments concerning the existence and structure of a human nature, her ethical concerns, as well as the scope and role, and her critique, of the natural sciences, especially developments in evolutionary biology.

According to Midgley,[30] *Beast and Man* is primarily about "the peculiarity and non-peculiarity of our species," a central theme that she has continued in subsequent writing. For example, in *Heart and Mind*, she dissected and criticized our tendencies to separate reason from emotion, as well as fact from value, mostly based on a misleading conception of the separation between mind and body (so-called Cartesian dualism). Then in *Wickedness*,[31] she dealt with the problem of evil, which by her own acknowledgment, she had previously overlooked.

---

[28] Mary Midgley, *The Owl of Minerva: A Memoir* (London: Routledge, 2005).

[29] Mary Midgley, ed., *Earthy Realism* (Exeter, UK: Imprint Academic, 2007).

[30] The source of my comments here is Mary Midgley herself, from a brief exchange of electronic mail between us on February 4, 2005; since that date, we have maintained a sporadic dialogue on diverse topics related to her work. I will have occasion to make reference to some elements of that discussion later on.

[31] Mary Midgley, *Wickedness: A Philosophical Essay* (reprint ed.; London: Routledge, 2001).

This last mentioned book is quite interesting from the perspective of the theologian. It is not theological in a technical sense, of course, but in a rather subtle way. Here she shares her insights on evil without resorting to the traditional theological question, the kind that asks, "Why does God permit bad things to happen?" Properly speaking, she frames the discussion squarely within the concern of a philosophical anthropology and psychology. As she has said, she deals with evil without ever having to mention or bring in God for anything. She believes that doing otherwise would be like sitting in judgment of God, but without any bearing yet on actually coming to terms with the problem of evil itself.

Finally, she pulled together the main arguments that originated in *Beast and Man* in *The Ethical Primate*,[32] where the issue of freedom, discussed within a solid understanding of Darwinian evolutionary theory and against the background of contemporary claims of biological determinism of all kinds, is central. She does it by first exposing all forms of reductionism, whether Cartesian dualism, Social Darwinism, or any tendencies to separatism and fragmentation, which impede a congruent, wholesome knowledge of our selves and, therefore, of the close relation, in this case, between ethics and consciousness. I personally consider it her best contribution to the general educated public, the one book of hers that everybody should definitely read.

Philosophy Is Public

Midgley's other books have branched out in different directions, though they are still "from the main stem,"[33] as she puts it. There are those books that deal with the nature and

---

[32] Mary Midgley, *The Ethical Primate: Humans, Freedom and Morality* (London: Routledge, 1994).

[33] A reference to the themes covered in her first book, *Beast and Man* (1978).

practice of philosophy itself and the ways in which people view its tasks. She is known for her use of the metaphor of "philosophical plumbing,"[34] a quite interesting image of her own creation, by which she explains the invisible or subtle presence of philosophical ideas in Western cultures.

As with ordinary plumbing, philosophical plumbing normally works quietly, doing its important job, and is not noticed until problems surface. Plumbing is never a neat system, never one simple structure, since it has been built through time, probably with disparate materials and planning. Ideas likewise do their quiet work of infrastructure until they stop functioning and therefore need fixing or replacement. At such a time, they can quickly make us aware of their existence by their own "stench," mess or malfunction.

Foundational ideas such as concepts, myths, and beliefs generate systems of thought and practice, which tend to create vast underlying connections, be they smooth or rough, and therefore are ideological structures. Systems of ideas and beliefs are generated through time, in an uneven way. Like plumbing, ideas are needed, and they go about doing their job whether we know or not how they work, and whether we like it or not. In any case, they can also cause lots of trouble.

There are also those books in which Midgley discourses on our relationship to other species, the animal side of the species barrier, and particularly about moral implications of our closeness to other animals,[35] especially the great apes. Of particular interest have been those books that deal with what science is: the nature, scope, and uses of a very impor-

---

[34] See, for example, Midgley's book *Utopias, Dolphins and Computers: Problems of Philosophical Plumbing* (London: Routledge, 1996), especially the first chapter, 1-14.

[35] Of particular importance to this issue is Midgley's book *Animals and Why They Matter* (Athens, Georgia: The University of Georgia, 1983).

tant and successful activity in our days. She has also penned a book and several articles on feminist perspectives on various issues.

Generally speaking, Midgley is ever concerned with what "passes under that name, science." She has been one of the major voices against a "pretentious science," whether in its "reductionistic" mode, viz. scientism, or in its "prophetic" or "salvific" dimension, viz. as a religion. In either case, we could argue, it is a kind of science that overstretches itself by violating its own methodological boundaries.[36] In general, and regardless of all claims to the contrary, science often mixes matters of fact with its own interpretation of meaning or significance. It hardly restricts itself to descriptive statements. It is probably incapable, like any other human endeavor, of limiting itself so strictly to separating facts from value judgments.

Although overtly polemical at times, Midgley has become something of a household name by her many publications in diverse media, including magazines, newspapers, and the Internet, as well as presentations on radio, television, and in other contexts. More recently, she was an honored guest of the Archbishop of Canterbury, Dr. Rowan Williams (b. 1950), as part of a series of "Conversations with the Archbishop" at St. Paul's Cathedral in London, where she displayed once more her ability to go directly to what really matters, the central concern or issue.[37]

---

[36] Many book essays and journal articles deal with the above issues and controversies, notably her fairly recent volume, *The Myths We Live By* (London: Routledge, 2003). For more information on Midgley's publications, see the attached bibliography.

[37] September 17, 2004 under the discussion topic "Humanity and Environment: Friends or Foes?" sponsored by the "St. Paul's Institute," in a panel with Dr. Ricardo Navarro from El Salvador, an engineer, a Christian layman, and environmentalist. Transcript of the presentation and conversation is available through St. Paul Cathedral's website at www.stpauls.co.uk

In academic circles, she has been known for heated debates on the proper understanding and consequences of developments in science, especially in genetics and evolutionary biology, with such figures as scientist and writer Richard Dawkins (b. 1941), and the late philosopher J. L. Mackie (1917–1981). In this latter case, it all began with the publication of Richard Dawkins' *The Selfish Gene*, in 1976, followed by an article by Mackie[38] drawing philosophical and ethical conclusions from it, which got a strong, almost angry reaction from Midgley. She responded to what she thought was most problematic in both of their writings, meaning, their support for a strong form of "psychological egoism" (or a least their lending themselves to that), with her own very controversial rebuttal.[39] The latter provoked three more polemical pieces, one by each of the participants in the dispute. The rest is history.[40]

The main accusation against Midgley's reaction was that she had misunderstood the science behind Dawkins' claims. Her critics' point, as far as the science is concerned, may have been essentially correct, but the problem lies in no small amount in Dawkins' ambiguous use of technical terms and in his writing style. As philosopher Gordon Graham has suggested, "Midgley ... took Dawkins too literally. ... Midgley's mistake, if that is what it was, is understandable, however, because Dawkins is himself inclined to draw

---

[38] See J. L. Mackie, "The Law of the Jungle: Moral Alternatives and Principles of Evolution" in *Philosophy* 53 (1978), 455–464.

[39] See Midgley's article "Gene-Juggling" in *Philosophy* 54 (1979), 439–458.

[40] For a very good telling of this story, see Ullica Segerstråle, *Defenders of the Truth: The Battle for Science in the Sociobiology Debate* (Oxford: Oxford UP, 2000), especially in chapter four, 71–74. For a slightly different view of the event, more from the perspective of a journalist, Andrew Brown, *The Darwinian Wars: The Scientific Battle for the Soul of Man* (London: Simon and Schuster, 1999), 83–93.

inferences, especially about human behavior, which suggest that the language of genetic selfishness and altruism is more than metaphor."[41] Those scientists, like Dawkins himself, who become good and popular at spreading and communicating their ideas to a wider lay audience, tend to overstate their case by becoming very much one-sided on the issue at hand. Instead of inviting further dialogue, they seem to alienate quite a few people.[42] The truth is that there is always a price to pay when anyone popularizes complex or technical ideas, scientific or otherwise. Dawkins has complained too many times of being misunderstood — by now, one may start to wonder what is going on.

As a moral philosopher, Midgley's passion is to understand both the conceptual and practical implications of dominant ideas. At a time when quite a few scientists have gone public, have come to demand much attention, and are treated like modern "sages," Midgley gives special attention to the kind of myths generated by their respective scientific worldviews.

In this regard, there seem to be two basic kinds of scientific endeavor performing the myth-making role. On the one hand, there is physics, which has long been considered supreme among the sciences and, together with mathematics, top in the so-called "hierarchy of knowledge." Therefore, it has become the measure that judges the accuracy of other scientific activities, above all in their cosmological manifestations. Moreover, as has often been asserted, the "Big Bang" theory or its equivalent provides the basic myth

---

[41] Gordon Graham, *Genes: A Philosophical Inquiry* (London: Routledge, 2002), 39. Later he adds that "Midgley's complaints against Dawkins are not without foundation", 44.

[42] For a balanced portrait of popular scientific writers, including Dawkins, see John Horgan, *The End of Science: Facing the Limits of Knowledge in the Twilight of the Scientific Age* (Reading, Mass.: Addison Wesley, 1996).

of origins to contemporary Western society. For Midgley, on the other hand, there is a more basic myth of origins, or at least one that has played more dramatically in the public arena and perception: evolutionary theory. She has called the latter "the creation-myth of our age."[43]

There is a long history of debates over evolution, due in no small part to its seemingly philosophical and social implications. That is particularly the case of its reductionistic offspring, for example, Social Darwinism and Spencerism. Developments of the latter kind have provoked fierce resistance among many parties to the use of evolution in the interpretation of human affairs. It has been, for example, identified with conservative politics, although this does not have to be the case, as Peter Singer has argued,[44] in my opinion persuasively. Nevertheless, evolution language has been used to argue a number of racist doctrines of physical or intellectual superiority by race, normally putting "White" on top, and as favoring intense social and economic competition. This state of affairs has been detrimental, time and again, to a more comprehensive awareness by the general public of basic biological knowledge, the kind that shapes all of us regardless of race or social status.

What Matters to Midgley

Acquaintance with the Midgleyan *corpus* of writings shows, amidst a panoply of interests and not a few informed

---

[43] For example, her opening remarks in "The Religion of Evolution", *Darwinism and Divinity: Essays on Evolution and Religious Belief* (Oxford: Blackwell, 1985), 155.

[44] Peter Singer, *A Darwinian Left: Politics, Evolution and Cooperation* (New Haven: Yale UP, 2000); he argues that Marx and Engels' belief in the malleable character of the human person, leading to the denial of a given innate human nature, has pushed many socialists to over-stress, to the point of exclusivity, social causes in the understanding of the human situation and predicament.

digressions, consistent strands of concern with how ideas, bad and good ones as well, shape the structure both of our thinking and also our behavior—not that she can be thought of as a "behaviorist" by any stretch of the imagination. Her concern is essentially philosophical and ethical, without trying to make a neat divide between the two. Of course, natural science, with its immense impact upon contemporary culture, in its multiple manifestations and with its myriad of activities, has attracted much of her attention. There is no way to overestimate its influence on our lives.

Therefore, what matters to Midgley above all is the role of science in human understanding. Science cannot explain, despite all its attempts, every sphere of human existence.[45] Midgley openly criticizes those who make exaggerated claims about science's powers. Promises are made that go beyond the obvious advantages of technological developments or advances in medicine. What she refers to is the promise of an ever better future, the teleological ethos, and the over-reliance on the human potential to re-make or re-generate the human self. This is science overstretching itself, aiming at reshaping the terms of other fields of inquiry, including those traditionally held by the so-called "human sciences." Again, her critique goes primarily against scientists' pretensions, those who deduce larger visions of the universe and of the place of humanity in it. Midgley is not bashing normal science *per se*.

However, the science that we have been sold, time and again, is not supposed to promise "heaven on earth" or the return to a "paradise lost." Being empirical and concerning itself strictly (and supposedly) to facts about the natural

---

[45] A case in point is that of Edward O. Wilson, *Consilience: The Unity of Knowledge* (New York: Knopf, 2003), which, despite his good points and understandable (for some, encouraging) attempts at using science to enhanced our vision of other fields of inquiry, ends up by making other forms of explanation subject or subservient to his brand of science.

world, the empirical realm, the coinage of new and almost superhuman values looks not just unreal but worrisome, to say the least. At the same time, in seeming contradiction to the actual tendencies referred to above, science insists on being reductive in character, always aiming at the simplest explanation, at what can be proven, experimented on, described in detail, and repeatable. In short, the realm of the experimental and observable, of what is measurable and quantifiable.

Others would include what philosopher Karl Popper (1902–1994) termed the "falsifiability" of science: that is what makes natural science truly empirical, and in contrast with many other activities which claim the label "science." The argument is that any of science's hypotheses and theories must be falsifiable in principle.[46] It means that controlled experiments can be enacted, not merely to prove a hypothesis or theory right, but attempting to prove it wrong, so to speak. Thus, any "science" that does not open itself to this kind of strict analysis and process of corroboration cannot be called such. It may be philosophy,[47] but not science. In the worst case, it may be religion or superstition—anything but science, properly speaking.

However, Popperian science is not the only understanding of what science is and how it comes about. The opposite view is that of Thomas Kuhn 1922–1996), who developed a

---

[46] The original source of these ideas and argumentation is from Karl Popper, *The Logic of Scientific Discovery* (reprint ed.; London: Routledge, 1992).

[47] This argument is actually being used in the case of "string theory": if a concrete experiment cannot be devised, if the theory as such cannot produce some kind of physical proof or experiential confirmation, even if the mathematics is beautiful, is it science? For more on this point, see the popular book (now also a documentary with commentaries by a few scientists) by Brian Greene, *The Elegant Universe: Superstrings, Hidden Dimensions, and the Quest for the Ultimate Theory* (New York: Vintage, 1999).

strong historical definition of the development of science and of its progress.[48] Kuhn's approach has been labeled "social constructive" by those who embraced it enthusiastically and have attempted to further expand his basic contention in different directions. New theories in what Kuhn terms "normal science" happen in a revolutionary way, when a previous theory or program has basically run its course owing to the internal contradictions that have been accumulating through time. Major shifts in science have occurred by leaps and bounds. New understandings of the universe, sometimes radically different, tend to replace more traditional ones.[49]

I wanted to refer to these two dominant views of what science is and does at this stage of our presentation because it will help later to clarify some of Midgley's criticism of science.

However, what mostly concerns me at this point is Midgley's assertion that, more than merely using a reductive approach, there is a kind of science that seems to be prone to scientistic ideology. It is the kind that attempts to explain human nature, and the structure of morality, with a simpler, more basic fact from science. The idea that we can find a single, fundamental explanation for complex phenomena is very much ingrained in the natural sciences, especially in physics and chemistry. However, when it comes to evolution of life forms, not to say then to human

---

[48] Thomas S. Kuhn, *The Structure of Scientific Revolutions*, 1962 (third ed.; Chicago: The University of Chicago, 1996).

[49] For what I think is a good, reliable presentation on this point of comparison, see Michael Ruse, *Mystery of Mysteries: Is Evolution a Social Construction?* (Cambridge, Mass.: Harvard UP, 1999), especially chapter one, 13–36; there he discusses what are considered the two main (or most influential, or at least talked-about, theories of science). A more recent volume dedicated to this issue is that of Steve Fuller, *Kuhn vs. Popper: The Struggle for the Soul of Science* (New York: Columbia UP, 2004).

motivation and conduct, the search for a single and simple explanation is but a dream, or rather an illusion.[50] Simplicity is a misguided approach, especially when applied to ethical, or bio-ethical, issues. Science does not have an easy solution, especially so in the case of problems that it has long excluded from its province.[51] In the past, it had excused itself from offering solutions to common everyday concerns.

Problems with Science

Midgley takes issue with the overuse of single principles in science, or in any other intellectual discipline, when they are invoked to explain too much. This is what she terms "reductionism." Reductionism happens when an explanation is stretched too thin, so to say. Moreover, reductionism in science is not science, but bad philosophy. It is the tendency to search for overall principles, to explain too much with too little. Reductionism is not logic—scientific or otherwise—but a kind of ideology, a habit linked with a variety of biased beliefs and attitudes. This has come about with utter disregard for the accumulated experience that is born out of centuries of deep reflection on human motives and activities. In short, it is a way of attaining knowledge that is oblivious to the kind of insights on human nature coming from the humanities.

A parallel problem, as Midgley sees it, is the appeal to simplicity itself as an ideal in science. The latter goes beyond the convenience of dividing a problem in science into smaller questions or parts for which simple, albeit temporary, solutions are sought. In science, simplicity is driven by the desire to exclude rival explanations or views from the

---

[50] For a thorough discussion on this point, see Mary Midgley, *The Ethical Primate: Humans, Freedom and Morality* (London: Routledge, 1996), 43ff.

[51] Midgley, *Evolution as a Religion: Strange Hopes and Stranger Fears* (revised ed.; London: Routledge, 2002), 62.

social imagination. Many times the goal is to alienate those views that are not strictly scientific in the narrowest sense. Religious understanding is a case in point. The public-savvy scientist considers it and treats it as superfluous, to say the least. As pointed out before, this is an example of the kind of ideological assumption that pushes scientific explanations beyond the lab and into the center of social life and over any other form of discourse.

However, notions about simplicity are not new to the intellectual landscape. I would like to suggest that it has a history of its own, not least in theological discourse. It is known that "simplicity," as applied to divine nature and activity, was a prime way of explaining divine nature among the medieval philosophers and theologians, Jewish, Muslim and Christian alike.[52]

A "simple" God is not a composite of elements or units or qualities. The Divine being is a whole, in which everything can only be understood as and only refer to a whole: the divinity itself. The world of experience, despite its multiplicity, has only one source, and everything refers back to the one principle of it all: God. Whether this God is personal, as was conceived by religious philosophers for the most part, or impersonal, considered by other philosophers as pure intellect, primary force, parallel to nature, or "nature" in the abstract, what mattered was that no imperfection or limitation could be attributed to this God. God is the one and only. There is no other God, and there is nothing like God. And yet, it is the one thought, source, principle,

---

[52] I have found David B. Burrell, *Knowing the Unknowable God: Ibn-Sina, Maimonides, Aquinas* (Notre Dame, Ind.: University of Notre Dame, 1986), especially 38–50, to be very useful in the understanding of this concept, especially as applied to God; Burrell however prefers the language of "simpleness" rather than "simplicity" since the latter term is often used in other contexts (for example, to speak of spiritual values or virtues) as well.

maybe "substance" that explains it all, either as origin or end of everything else that is the case.

Reductionism in science and simplicity of explanation—whether it is the "all in the genes" explanation in evolutionary psychology, natural selection in biology, a "theory of everything" in physics, or a configuration of atoms and molecules in biochemistry—seem to me to be theologically or at least religiously motivated. What I mean is that I see it as a kind of "carry over" from centuries past. I cannot help but think of medieval philosophers and their insistence on and defense of divine simplicity as a unitary and metaphysical principle, since the aim after all was to explain all of reality. Now this unitary ideal is applied to nature, the one mathematical equation or law of nature, the development of humanity, or what makes us human in the last analysis.

According to Midgley, the link to monotheism behind the drive to reductive or simple explanations may be right. People are so desperate for simplicity in a confusing world that they also "try to simplify God continually, control him or her, and shape him or her to fit their requirement."[53] The point that I am trying to make is not whether this is right or wrong—I actually believe strongly in the independence of scientific research, unheeded by previous ideological commitments, especially of a negative kind. What I do want to say, along with Midgley, is that when it comes to science, it is emerging today that the seventeenth century's bold assumption of simplicity in nature was not justified. Some underlying simplicities can indeed be found beneath complexity. But further complexity often emerges beyond them. There is no guarantee that we shall ever be able to get a "final answer" and, according to Midgley, nor do we need one.

---

[53] As mentioned before, the source of Midgley's comments quoted here is our personal exchange through e-mail back in February 2005 and in subsequent communications since then.

More matters to Midgley than coming to terms with science. Philosophy has played a significant role, probably contrary to public opinion, in shaping our conceptions and attitudes about the world, including the whole of nature and especially animals, and the human community. She advocates reading philosophy as widely as possible. Again, we have her contention about dominant ideas as functioning systems, or as "philosophical plumbing." For any ideology, scientific or otherwise, there are unspoken assumptions. More than that, there are veiled commitments, political or otherwise. Awareness and self-criticism can be very much helped by philosophical analysis. Philosophy belongs into daily life, in the free and open market of ideas.

Midgley has advocated, together with her husband Geoffrey — and they are not alone — what they call "applied philosophy." This is basically a way of applying philosophical concerns to real life, to real life issues. Any of her books abound in examples on how to philosophize "in the streets" and not just in the academy. She has shown the "how to do it" by example, by dealing with topics as varied as human relations and psychology, relation of humans to animals, animal rights, issues confronting women and feminism, religion and science, bio-technology, health care, education, politics, war on terrorism, just to name a few important issues.

Some conception of the importance of the "inner life," or what religious thinkers sometimes call "spirituality," also matters to Midgley. As for Iris Murdoch, so for Midgley the inner life of motives, thoughts, intuition, and even instincts, are essential to our power over action. It is a source of the moral life. Many object that the inner life is not really "quantifiable", or even "knowledgeable" since there is no firm manner of corroborating what others think or feel, properly speaking. But for Midgley, this is not a problem of either philosophical or psychological "solipsism." The difficulty lies rather in our tendency to split thought from action, reason from feeling, inner from outer living. These are all quite intermingled, and quite difficult to distinguish clearly or separate one from the other, despite the fact that so much philosophical capital has been spent in doing precisely that.

Dualisms of many kinds are metaphors, not direct descriptions of "what is real." We use them all the time, probably unconsciously. For example, the dualisms of mind and body (known philosophically as "Cartesian dualism"), heart and mind, humans and nature, and the individual and society, among others, are frequently either invoked or just assumed. However, these are rather metaphors, albeit strong ones, which are used to build ideas, systems of ideas and ideologies. And metaphors are used to build myths, the dominant reasons for doing this rather than that, and for explanations of all kinds. These are the ways in which we approximate an understanding of our own "reality," social or metaphysical.

It is essential to avoid setting "polarities" of the above kind against each other in a sort of "win-lose" match.[54] This can be done without necessarily eliminating conceptual distinctions, which are akin to how our minds work. According to Midgley, however, what we need is to put them into a larger context, as pairs of interplaying concepts or images, where they can relate to each other — in a "moderate dialectic" — instead of submitting one concept (or task) to the other and without finally getting rid of either side of the binomial. The most important thing here is to avoid extremism. When getting out of hand, it is such extremism that, with a conceptual language at its disposal and good and common images to support in a public ready to be persuaded, can become a myth for a whole group or society.

---

[54] On this point, see the discussion on some of Midgley's key ideas in James M. Gustafson, "Scientific Dreamers and Religious Speculation," *Christian Century* (March 10, 1993), 269–274.

One current myth is the generally accepted opinion that there is a view from inside, which is not scientific, and a view from the outside, which is scientific, actually science itself. This is born out of scientists' desires to attend only to the "facts" of nature without any intervention from preconceptions, emotions, or bias, in short from "subjective values." Even the notion of interpretation, either in the form of one possible interpretation among many, or of equally competing alternatives, seems anathema to some scientists. According to Midgley, noble as the attempt may sound, the goal of disassociating fact from belief rests on an illusion. What happens more often than not is that one kind of belief is substituted for another. And when people are blind and therefore unaware of it, the results are not good. Science is offered, not just as *a* reasonable solution to some problems but also as *the* solution to all of our problems. "Scientific" becomes a code word and thus synonymous with concrete, rational, and truthful.

The opposite is then any explanation that is merely believed, intuited, or abstract. But why, Midgley asks, is belief always thought to show weakness? Such an attitude is the prejudice of a culture, one not just deeply influenced by science, but certain of itself as being truly "scientific."

In truth, the difficulty in trying to make such a clean cut between fact and value, objective knowledge and subjective experience, lies in human nature itself; it has changed with the arrival of modern science. Thoughts and emotions, reason and intuition, facts and values are all ways in which we grasp the world, by which we make sense out of the myriad of experiences. Obviously, we need all of the above. They work together. These are all functions of human cognition.[55] We do have to submit our emotions to rational analysis, but also allow deep-seated emotions and nonconformity to

---

[55] Midgley, *Heart and Mind*, especially the introductory chapter, 1–38.

have a say. Emotions have a discriminatory power. Therefore, emotions should not be dismissed offhand. Reasoning has an emotional response; emotions can have reasons intrinsic to them.

We need to learn to sort out whatever contradictions may confront us. Which is not to say that we can do without one or the other—we cannot get rid either of "heart" or "mind." It will not do for us to reject either of those two. Besides, each term stands for a number of cognitive functions in the human person, such as perception, intuition, deduction, intentions, etc. These all come together in order to organize our experience and, as often as possible, become unavoidable steps in the creation of knowledge.

Individualism Is Not the Solution

There is a modern bias for individualism, basically towards competition as central to human nature. And now we have the philosophical and psychological theories to go along with this.[56] Individualism is one of our most forceful motivations in contemporary society.

On the one hand, we worship individual success, the kind that leads to the greed called materialism. Most of the goods sought are used to enhance individual status, rather than to merely satisfy essential needs. On the other hand, it is believed that the tendency toward betterment and therefore toward progress is natural, meaning that we have been "hardwired" by evolution to look after our own selfish aims.

---

[56] Midgley has written about her concern with individualism in many of her publications. For our brief presentation here, I rely for the most part in her essay "Toward a New Understanding of Human Nature: The Limits of Individualism," *How Humans Adapt: A Biocultural Odyssey*, ed. Donald J. Ortner (Washington, D. C.: Smithsonian Institution, 1983), 517–533.

Midgley is convinced that, contrary to current opinion, these beliefs have no ground in Darwinian evolutionary theory. Ours is not a species that thrives only on selfish motives and instincts, or is driven exclusively by competition. We have been able historically to detect egoism and know it for what it is, and can also be very critical about it, while trying to avoid behaving this way. Individualism is neither the only nor the final answer to the human situation and predicament.

However, not all individualism is negative. Since the Renaissance, and especially with the development of modern liberalism, there have also been many attempts to liberate individuals from the undue pressure of their social background. The recent history of many societies shows the myriad of efforts to teach and support individuals to act freely and stand on their own feet, to take responsibility by themselves and for themselves, and to claim their inalienable rights. All that said, liberal political theory also taught that individuals are like atoms, autonomous units, persons who are able to decide to come together for mutual benefit under a "social contract."[57]

This idea of a single, contracted individual has been considerably strengthened since the European Enlightenment. However, according to Midgley, contractual thinking is limited. For one it does not extend the same rights to other fellow humans in far regions of the planet, considered for the most part as either "uncivilized" or "underdeveloped". Moreover, there seems to be no space in the contract for non-human animals or for the environment. And it cannot provide the latter, because it is flawed. It does not have a notion of "right," not even a word that could be applied to other living creatures or to the earth as a whole. This is where exclusive self-interest stops helping and starts hurting in its stead.

---

[57] Midgley, 519–520.

Some scientists, for example ecologists and ethologists,[58] have shown us something different: the mutual dependence between humans and their environment, and the close familiarity between them and other animals and organisms. Also, social scientists and anthropologists have pointed out the importance of communal support and values, our deep dependence on our social background, and the continuity of cultural demands.[59] These are all vast and complex systems that made us who we are and still nurture us. No person is self-made. We are all products of the complex interaction of biology and the environment, instincts and ideas, or brains and cultures.

It is also the case that the new vision takes time to sink in—besides we have a tendency to revert to old views and values almost naturally. Change is always hard to do, and changing minds and attitudes is even harder. Individualism, in all its manifestations—whether physically or spiritually conceived—retains a strong influence in many theories: scientific, economic, religious, etc.

Of course, argues Midgley, human beings are distinct individuals.[60] But humans are also an integral part of this planet and have a common fate with it as well as with many of its creatures. Harmful changes to the environment can change life as we know it. Humans are social beings, dependent on their social life as well as on their biology. These are both rather fragile. More individualism is surely

---

[58] "Ethology" is basically the scientific study of animal behavior, especially as it occurs in a natural environment, with special attention to its evolutionary and ecological aspects. As a discipline, it has been deeply shaped through the work of the German Konrad Lorenz (1903–1989) and the Ducth Nikolaas Tinbergen (1907–1988).

[59] Midgley, "Toward a New Understanding of Human Nature," 521.

[60] Midgley, *Evolution as a Religion*, 169.

not an alternative to this fragility—individualism can actually make things worse for all.

As Midgley says, it is not by wrecking the biosphere,[61] even if we think that for that to happen takes a long time, nor by disrupting our communal life that there will be found solutions to many of our most pressing challenges. It is seriously difficult to have to choose between, for example, "*conserving* the environment for human use and *preserving* it for its own sake."[62] We should not have to apply an either-or approach to our search for solutions to the situations that threaten us. An either-or solution is the type that the modern mind has us accustomed to, but it is not necessary nor will it ever work properly.

Why Midgley Matters

A good example of the outcome of this conflict of perspectives, and the idea that by splitting approaches we reach a solution, is the science of evolution, as it has been defended and practiced for so long now. For example, attributing motives to evolutionary functions, and seeing them as in competition to ordinary motives, have been major problems in dealing with explanations for human behavior coming out of evolutionary psychology, or from some forms of genetic determinism. These are examples of a kind of scientific practice and discourse that tends to use partial causes, which it studies with determined attention to detail, as ultimate causes of things, whatever that thing, activity, or behavior, is.

According to Midgley, a science of ultimate causes is bad science. Since science is at its best studying partial causes,

---

[61] Midgley, "Toward a New Understanding of Human Nature," 532.

[62] Midgley, 532; emphasis is hers. Here she summarizes John Passmore's posing of this particular dilemma in his book *Man's Responsibility for Nature* (London: Duckworth, 1974).

when it does otherwise, it creates confusion. Moreover, it is the kind of science that begins to ring of religious overtones! When science gets into the realms of, for example, immortality, human destiny, and the meaning of life, it has already gone beyond itself.[63]

Which is not to say that we can practice science detached from any other concerns. Our theoretical curiosity is not detached from the rest of life, and science does not work in a vacuum anymore than any other activity or form of thought. As with any other human intellectual activity, science also provides information that fills in world pictures.[64] In this sense, science is a kind of map.

Midgley likes the metaphor of different maps to describe a territory. The image or model that she has in mind is that of the relation between different maps of the same territory as in an atlas.[65] We could use a map of rivers or mountains, roads or landmarks, population density, etc. To think that we only need one map to get to know well a place or territory is ludicrous. No map can show everything that there is to know. Each map, even if needed at different moments or for different purposes, is equally valid and useful in its own way. All maps convey information, and answer different questions, and none can be sufficient for all times and purposes, nor suffice alone. The point is, that we can have different kinds of explanation without necessarily being in competition.

Someone who does not have a map or frame of reference may accumulate facts or information for no use or good

---

[63] Midgley, *Evolution as a Religion*, 1.

[64] Midgley, 2.

[65] See for example Midgley's description of this model in her article "Pluralism: The Many Maps Model", *Philosophy Now* 35 (2002), 10–11. Her multiple contributions to this magazine are good examples of her concern with the popularization and critique of important philosophical ideas in different media for their broader appeal.

reason—it is an "idiot's task." All that said, every map points to a whole, its general frame of reference. The latter is shared with other, though different, maps.

Midgley's work itself is a kind of map, or rather a series of maps that could be used interposed. First, there is a map that is conceptual: it unveils systems of thought or ideas. To use her own analogy, one to which we referred before, it is about "philosophical plumbing." In addition, there is the map of human relations, whether it is about relations with other humans or the non-human animals. And, finally, there is also the map of beliefs, which are (at least for Midgley) thoroughly human in both origin and character. I think that these are the relevant ones in order to understand her approach. These are also useful for anyone with equal or similar concerns.

To tell the truth, Midgley's attention has not really focused on theology. It is not because she feels it is a waste of time. On the contrary, she has at times acknowledged theology's historical experience, rigor, and complexity. It is obvious to her that theology has a long history, and that much has happened in it—even if some scientists or scholars have not noticed or refused to notice—since Galileo. Nevertheless, if she had given theology its due attention, she would surely have submitted it, as James Gustafson has said, to the same kind of criticism and witticism that she has used for other fields of enquiry—especially when "big ideas" are breathed in and out. One case in particular is revealing.

In a response to another author's summary of [protestant theologian Karl] Barth on creation, Midgley wrote:

This sounds to me quite simply mad, and mad for entirely traditional reasons. Where, after all, was Karl Barth when the Lord laid the foundations of the earth?[66]

In any case, and going to the previous point, we may disagree on the origin and function of certain ideas or practices, whether those are, say, any of the "big ideas" by which she insists that society functions (for example, the "social contract") or any other specific beliefs about which people feel especially passionate. But we can only agree with Midgley's conviction that these all play a forming role. Their impact is too great as to be ignored or dismissed. Their role goes beyond religion and into science and philosophy themselves, where we normally did not expect to find them.

In order to get a good glimpse at shaping ideas or myths, and especially in the case of the sciences, which are so culturally predominant today, we need Midgley. This is among other reasons why Midgley matters.

---

[66] Gustafson, "Scientific Dreamers and Religious Speculation," 270. Midgley's language is a reference to a biblical passage in the Book of Job, chapter 38, verse 4a, where God addresses Job in this manner: "Where were you when I laid the foundation of the earth?" (text from the New Revised Standard Version).

Chapter Two

What Concerns Midgley

As a philosopher, Mary Midgley takes all of history, especially the history of ideas, very seriously. Intellectual history is a major source for her analysis and criticism of contemporary movements and philosophical systems. Scholars often criticize Midgley for dealing so much with the past's foregone ideas or extinct movements. One frequently cited example is the many pages that she has dedicated to the now faded programme of "behaviourism" *à la* B. F. Skinner (1904–1990)[1] et al. Another example is the many times that she goes back to Francis Bacon (1561–1626)[2] in order to explain a peculiarity of the way in which

---

[1] Skinner is considered a major force in the so-called scientific study of behavior, animal and human. His lab experiments in the modification of behavior were and remain controversial to this day. The author of several books expounding the results of his research and ideas, he has been also strongly criticized for his belief that animal behavior is basically malleable and therefore programmable, to the exclusion of naturally given behavioral and traits and character. Among his most popular works are *Walden Two* (1948), *About Behaviorism* (1969), *Beyond Freedom and Dignity* (1971).

[2] Bacon was a philosopher and statesman in the reigns of Elizabeth I and her successor in England. He is widely known for his defense and exposition of the nature of the scientific method.

society, in her view, conceives of science and of the role of science in society. I believe that one thing that the critics miss is the very nature of philosophy as a discipline, and of Midgley's understanding of the impact and multifaceted role of ideas, old and new, in society.

Basically, Midgley is in dialogue with philosophy's past and present. As a matter of fact, to distinguish neatly between those two moments in time does not do justice to the complexity of the issue at hand. I am reminded of the discussion, back in the late 1970s, about the relationship of philosophy with its past, and the role of a history of philosophy within the field.[3] It so happens that philosophy, unlike, say, science, has a unique relation with its past, where the latter is never "out of sight", so to speak. Philosophical ideas and forms of thought, with very few exceptions, do not just disappear for good; they linger. If anything, science and the technology change, cultures evolve in different ways, but philosophical questions and attempts to answer those questions stay around, albeit in different dress.

Normally science students dealing with physics, biology, or chemistry, unless they are specializing in the history of the discipline in question, do not really need to know previous but discarded theories. The old theories are inconsequential for the task of doing their respective science successfully today. It is not so in philosophy. In what sense have Plato, Aristotle, and Descartes been overcome, proven wrong, or defeated once and for all? Can we really understand how many forms of realism there are today without the development of a fruitfully diverse, multifaceted, com-

---

Among his main works are the *Novum Organon* (1620) and *The Advancement of Learning* (1623).

[3] The volume that I have in mind is that by Jonathan Rée et al., *Philosophy and Its Past* (Atlantic Highlands, N.J.: Humanities P, 1978), which was a required reading in a seminar on Hegel's "Lectures on the History of Philosophy."

plex idea about the world and reality? Have ethical questions and concerns changed so much since Socrates' time that they are no longer recognizable as such? Is philosophical atomism really new?

It is intrinsic to Midgley's job as a philosopher to unveil the deep connections among many ideas, past and present, and to unveil the role that these play in our own context, in the way that we as society view the world around us. Part of the beauty of Midgley's writings is precisely that ongoing conversation with a past that is still so much part of the present by way of the renewal of ideas. She engages the likes of Hobbes, Descartes, and Hume, or Plato and Aristotle, with certain ease. I find her approach more than merely instructive: it is also relevant to the task of exploring the origin, development, and especially, the resilience of ideas. Those ideas and systems of thought are not peculiar to a philosophy properly speaking—I say this in consideration for the many who restrict the history of philosophy exclusively to the "great ideas" and especially to the "great men" (sic).

However, Midgley finds relevant ideas in every discipline, in cosmology and in psychology, religion and theology, science, and even art. Ideas, not to say ideologies, cross many boundaries; and sometimes appear even in the most unexpected places—just another case of what Midgley calls "philosophical plumbing." Most peculiarly, she finds that it is now that scientific ideas percolate to many other fields of knowledge.[4]

Some themes have been visited by Midgley on a regular basis. These are the ones that have, in her appraisal, the most impact, whether by danger or promise, in current attempts to come to terms with the world as we think we know it. At times she takes sides, and at times tries to remain neutral, or at least to leave it to the reader to decide on his or her

---

[4]   See Midgley, *The Myths We Live By* (London: Routledge, 2003), especially chapter four, "Thought Has Many Forms," 21-28.

own—although not without previous warning of the possible consequences. Inconsequential or inconsistent ideas do not really bother her, but then there are those that manifest a life of their own. Some visions are just more powerful or longer lasting than others. And she wants to know why it is so.

Since the nineteenth century one such big idea has grown ever more powerful and has metamorphosed in ways that have brought it up to the point of becoming the predominant "myth-making" of our times: not so much "Big Bang" cosmology but rather Darwinian evolution.[5] According to philosopher Daniel Dennett, Darwin's idea is the single best one that anyone ever had, ahead of Newton, Einstein and everyone else. It has a unifying power unlike any other idea. It has unified "the realms of life, meaning, and purpose with the realm of space and time, cause and effect, mechanism and physical law."[6] For others, the theory of evolution "has become the symbol and sign of a type of thought which affirms life, which inserts human beings into the stream of all life and puts all life into the context of the all-embracing cosmic reality. It is ... a component of the Western view of the world and of life which is accepted everywhere."[7] Many seem to agree on this assessment: evolution is a more funda-

---

[5] Midgley is not by any means a lonely "voice in the desert" in this regard. See the essays in Bas van Iersel et al, eds., *Evolution and Faith*, Concilium 2000/01 (London: SCM P, 2000). However, I find Midgley's exposition normally clearer and more compelling in the use of concrete references to the work of many authors that she engages than those of many other writers. No wonder she has been called repeatedly "a philosopher that the average person can understand." Moreover, she has her own particular contribution in regard to how evolution doubles as metaphysics and religion but without proper acknowledgement, something which I hope to make clear in this work.

[6] Daniel Dennett, *Darwin's Dangerous Idea: Evolution and the Meanings of Life* (New York: Simon & Schuster, 1995), 21.

[7] Bas van Iersel et al., eds., *Evolution and Faith* (London: SCM, 2000), 24–25.

mental myth of origins and, precisely for this reason, a more controversial one too. It makes many people feel uncomfortable. This situation has not changed much since Darwin's day.

What with Evolution then?

According to Midgley, evolution works as a religion. In religious thinking, the inexplicable and the incomprehensible nature of things holds a promise of better existence; it offers salvation, or a fulfillment of sorts, while providing a sensible picture of the universe. And so it happens with a kind of science, one that promises more than it can actually offer or deliver.[8] A science that promises immortality, be it through cosmological exploration, the reproduction of life, or bio-technological enhancement of human capacities tries to do more that it can or should do.[9] This is not purely and simply "science"; this is rather science as "salvation."

In the case of evolution (or, more appropriately, "evolutionism"[10]), it pretends to provide a sense of wholeness together with an understanding of origin as well as purpose. It speaks of progress, while it promises survival, at

[8] This argument is fully developed in Midgley's *Science as Salvation: A Modern Myth and its Meanings*, 1992 (London: Routledge, 1994); this book is based on her Gifford Lectures at the University of Edinburgh in 1990.

[9] Midgley's writings abound with example of such "scientific" proposals; some will be mentioned later on. However, just for a recent token of such "scientific optimism" I want to refer the reader to Naam Ramez, *More than Human: Embracing the Promise of Biological Enhancement* (New York: Broadway, 2005). For an open yet balanced approach to the role of technology in human development, see Philip Hefner, *Technology and Human Becoming* (Minneapolis: Fortress, 2003).

[10] The way of naming one form or another (say between "evolution", the "theory of evolution", and "evolutionism") is helpfully explained by Michael Ruse, *The Evolution-Creation Struggle* (Cambridge, Mass.: Harvard UP, 2005), 4.

least for the best adapted (or the fittest, in Spencerian terminology). It tries to unify what other forms of science (e.g. physics) do not: facts and values, albeit in a limited, materialistic way. Evolution, in other words, is not just science. It is also a myth, a designation that is not intended by Midgley to mean a lie, or something necessarily untrue or fictitious, about human origins.[11] A myth tells a story, often about origins, precedence, and maybe even purpose. Some myths aim to be comprehensive, providing a worldview. Others just intend to supply guidance on some aspect of life. In the case of evolution, I think that it means different things for different people: it is philosophy, belief, and ethics as well.

Evolutionary thinking has influenced, and in some cases created, a number of disciplines or new approaches within established fields of human inquiry. Whether in psychology, consciousness studies, cosmology, social analysis, ethics or economics, it is a common place nowadays to see new understandings that draw on some aspects of Darwinian evolution and transpose those to other fields, while coining new "scientific languages." It is also behind the doctrine of progress: the strong belief that the advancement and fate of humanity is inevitable and, at the same time, inextricably related to science. They go hand in hand. In most cases nowadays, explanations by evolution seem everywhere present.

Questions of meaning probe our ability to find any such thing. But the point here is, that it has been left to science to deal with meaning too. And this is a problem. The answer, however, is not to revert to old dichotomies, like those that stress the strict separation between "facts and values," or between "body and mind," or "science and the humanities." For Midgley, no search for a single explanation or defining principle for meaning in human nature and life is a

---

[11] Midgley's main work on this issue is *Evolution as a Religion: Strange Hopes and Stranger Fears* (revised ed.; London: Routledge, 2002).

disinterested one. What we need, however, is a plurality of approaches, scientific or otherwise, a systematic use of different ways of thinking in different contexts to answer different questions, where we cannot do either without science, religion, or other forms of knowledge.

In line with Christian theologian Jürgen Moltmann,[12] we can say that insights from both science and religion are needed in our "way to wisdom." No one explanation, doctrine or approach will ever be able to offer an understanding of the whole in all its complexity. Neither science nor religion should necessarily aim at the least and simplest. The latter has undoubtedly a role in the process of communicating with the larger public. But it could be rather detrimental if either were used to eliminate any thoughts about the complexity and elusiveness of reality.

The Epistemological and the Ethical

Questions about the shape and meaning of reality lead inevitably to questions about epistemology, a fancy word to speak about the way we know and knowing how we know and what it is that we know. Epistemology is about "knowing our knowing," as Catholic theologian Ivone Gebara likes to say.[13] It is not just about bites of information, or

---

[12] Jürgen Moltmann, *Science and Wisdom*, trans. Margaret Kohl (Minneapolis: Fortress, 2003).

[13] Ivone Gebara, *Longing for Running Water: Ecofeminism and Liberation*, trans. David Molineaux (Minneapolis: Fortress, 1999); see chapter one, 1–65. Gebara is a preeminent Brazilian Liberation theologian, known for her phenomenological approach in dealing with issues on ethics and the environment, as well as feminist critiques of social realities. But she has also tackled questions in systematic theology, from the doctrine of God, through Christology, to Marian doctrine. Questions around Roman Catholic teaching have put her in conflict with the Vatican, including the imposition of a period of silence and further study in order to "correct" her doctrinal views.

collection of data. It is also about feelings, experience, and intuition. Our knowing needs and uses as many epistemological "tools" as it can fashion. This is how we have evolved as knowing beings, as a species—not that we are the only ones, or that that is all we do.

More importantly, our knowing has ethical implications. What we know affects our response. That is, what we know affects the quality of our actions. There is an ethical dimension to human knowledge. As Mary Midgley suggests, any attempt at establishing and abiding by a specified intellectual explanation leads to moral questions.[14] Moreover, morality is essential to human life and wellness. It cannot be reduced to a mere tool—egoistic or whatever—of evolutionary processes. Which is not to say that it has not evolved with us.

Midgley argues that ethics is quite essential to human development, and not to be seen as extra baggage that we are carrying merely by custom or tradition. Therefore, she needs to raise the question about the origins of ethics, which is to say, the origins of social behavior and communal life. Midgley believes that she finds the key to the origins of ethics in human evolution, through a very long history which moves from the development of social instinct among the ancestors of *Homo sapiens*, to the inevitable resurgence of conflict, to the necessary development of the resolution of conflicts. Therefore, both for the study and development of moral principles, we need more than just religion or metaphysics: we need biology as well.

Certain kinds of knowledge are compelling. We cannot, and are not allowed to, remain neutral or unresponsive. Even such a stance as not getting involved in something is already a response of a sort: it carries a decision, with some

---

[14] See Midgley, *The Ethical Primate: Humans, Freedom and Morality* (London: Routledge, 1996), 73; especially on the matter of facts, which are not necessarily value-free, but are needed nevertheless in making moral choices.

ethical value.[15] This is inevitable. We are built (or hard-wired, if you wish) that way. There is no guarantee that the actual conviction or outcome will necessarily be moral in the good sense of the term, meaning, that it will not necessarily facilitate the good or a higher good, or the good for all. However, for people who care about the fate of others, of communities around the world, and for those who care about the environment where human affairs evolve, human solidarity is a prime response.

For someone like Midgley, our ethical response should not rule out offhand, as happens so often, non-human animals. She has been involved for years with animal rights and conservation movements in England and other places. Midgley has written extensively about this topic.[16] She has lent her name to multiple causes involving advocacy for animals' well-being. For her, the reasons for her commitment are deeply philosophical as well as moral. And, incidentally, her level of involvement and commitment to animal conservation and respect for life also convinced her a long time ago to become a vegetarian.

For the most part, philosophers have neglected the animal question. Part of the problem is due to the difficulty of how to fit this concern into our ethical systems. Disregard for animals' well-being, not to say animal rights, has a long history in Western thought. Thinking about animals has not

---

[15] On this point, I draw partly from the excellent treatment of the subject by Mary M. Solberg, *Compelling Knowledge: A Feminist Proposal for an Epistemology of the Cross* (New York: SUNY, 1997), 1–8. Solberg is a Lutheran theologian who worked in El Salvador during the civil war years, and who developed her ethical theory by contrasting people's moral responses to war in Central America and in the USA; she found different ways of getting to know and responding to a critical situation.

[16] Midgley's major statement in this regard is her book *Animals and Why They Matter* (Athens, Georgia: The University of Georgia, 1983).

been deemed worth the effort, unless it is useful to define human uniqueness. Christian thought, with a few exceptions (St. Francis, for example), has denied souls to animals and, therefore, excluded them from further consideration.[17]

Nevertheless, Midgley's statements cannot necessarily mean that no thought was given to defining animal nature. In fact, the need to come to terms with what exactly in the human owes its being to the rational and the spiritual (to the soul, in brief) and how much is purely "flesh," or animal instinct, was regarded as an important study by those interested in answering the question as to what kind of bodily resurrection Christians were going to enjoy. The ensuing debates, about whether nutrition or reproduction—more exactly, intestines and sexual organs—were still going to be of any use in the heavenly paradise, provoked a number of creatively interesting responses to the issue at hand.[18]

Again, it was a matter of coming to terms with human nature and its animal tendencies, which at least some theologians thought obvious in the first place.

Since the Renaissance, a strong humanism, or *reductive humanism* as Midgley likes to call it,[19] has had no place for animals, except to say how low in the "scale of being" they are in comparison with humans. Moreover, it has denied animals any dignity, considered to be essentially a human attribute, and spoken of them as subjected to human power and control. With the advent of modernity (post-seventeenth century), humans have seen themselves not just

---

[17] Midgley, 10.

[18] For an educating (yet entertaining) look at these issues, see the beautifully written book by Italian philosopher Giorgio Agambe, *The Open: Man and Animal* (Stanford, Cal.: Stanford UP, 2004), especially chapter five, "Physiology of the Blessed," 17–19.

[19] See, for example, Midgley's essay "The Paradox of Humanism" in the collection *James M. Gustafson's Theocentric Ethics: Interpretations and Assessments*, edited by Harlan R. Beckley and Charles M. Swezey (Macon: Mercer UP, 1988), 187–199.

above but moreover in competition with or against nature. We have not accorded a much-needed respect to non-human animals. There has been a call for complete dominion over nature and the animal realm. Any link or parentage between us and other animals has thus been emphatically denied.

During the Enlightenment, despite the efforts of a few in the opposite direction, the exaltation of reason and those faculties deemed rational as distinctive of humans, led to further neglect of animals. According to Midgley, reason came to play the same role as the soul in Christian thought:[20] the operator in charge of controlling all other functions, from thinking to feelings as well as the will behind motion. Descartes however had already denied any form of consciousness to animals, considering them "automata", basically machines, creatures with mechanical structures, strictly led by and reacting to mechanical forces.

Today the common view seems to be that, in general, humans have essentially the faculties that other animals have but with unique, crucial and developed differences or additions.[21] We humans do not have just one difference from animals: we have differences from different animals, both in degree and quality, as we have many similarities with many of them. The increasing conviction — especially thanks to the work of ethologists in general and primatologists in particular — is that we cannot attend to the meaning of human nature without closer attention to animals. This new attitude is due, in no small part, to a new appreciation of Darwin's contributions to the study of animals, animal behavior and emotions, and their environ-

---

[20] Midgley, *Animals and Why They Matter*, 11.
[21] Midgley, 12.

ments.[22] With the help of contemporary mass media, Midgley argues, Darwin is finally getting through. As Midgley herself states it,

> For the first time in civilized history, people who were interested in animals because they wanted to understand them, rather than just to eat or yoke or shoot or stuff them, have been able to advance that understanding by scientific means, and to convey some of it to the inquisitive public. Animals have to some extent come off the page. With the bizarre assistance of TV, Darwin is at last getting through. Town-dwellers are beginning to notice the biosphere.[23]

Midgley finds hope for the animal world in the media's attention to their critical predicament. This is something that is probably working better, though not exclusively, in Europe than in the United States of America, from what I can tell.

Is There a Mind-Body Problem?

For Midgley, one of the major obstacles to thinking clearly about the world and our place in it has to do with the so-called "Cartesian dualism" or the "mind and body" problem. Of course, the issue is never trivial since it is deeply rooted in self-conception, that is, the way we think of ourselves in relation to the world. For centuries, philoso-

---

[22] No doubt this attribution to Darwin may be cause for controversy. Darwin was not the first or only one to take animals seriously and examine them closely. All that said, what I believe Midgley has in mind is the fact that Darwin produced one of the most substantial studies based on close and detail observation and analysis of animal conduct under different environments; I am referring now his monumental work *The Expressions of Emotions on Man and Animal* (reprint ed.; Oxford: Oxford UP, 1998), originally published in 1872 with a second edition in 1889, after Darwin's death. The other only good example that comes to mind is that of Aristotle with his work on biology.

[23] Midgley, *Animals and Why They* Matter, 14.

phers have wondered about the way in which we relate our inner and outer realities. They seem to have concluded that, for the most part, we not only think differently about those two kinds of perceptions but that, worse yet, there seems to be an unbridgeable gap between the two and that we ourselves are split in that way.

Are the inner and the outer lives two separate items, or two aspects of a whole person? According to Midgley, if we think of mind and matter in either way, we are thinking of mere abstractions.[24] René Descartes saw them as incompatible. In his time, this separation was intended to keep different disciplines, for example, physics and metaphysics, separate, each minding its own business. He thought it necessary in order to protect physics, which was at the time the new science and, therefore, in need of autonomy to grow. However, it was also an attempt to establish the priority and, above all, superiority of mind over body, thought over matter, and reason over feeling.

Mind and matter came to be considered as two separate things, two very different kinds of stuff. In England, for instance, this basic dichotomy ruled unquestioned well into the twentieth century, when philosopher Gilbert Ryle published his influential treatise on the concept of the mind,[25] in which he persuaded other philosophers to stop talking of a "ghost in the machine," the usual metaphor. Unfortunately, this mode of thought prevailed among the general public. Instead of keeping both concepts together, and working with both as in an intimate relationship, the tendency to treat only one to the detriment of the other, or to try to

---

[24] For my summary of these concepts and debates here, I draw primarily from Mary Midgley's article, "Souls, Minds, Bodies, and Planets," *Philosophy Now* 47 (2004), 33-35.

[25] Gilbert Ryle, *The Concept of the Mind*, 1949 (reprint ed.; Chicago: University of Chicago, 1984), especially the first chapter, "Descartes' Myth," 11-24, where Ryle defines the basic problem.

explain the whole by one of two extremes, has been rather common.

During the twentieth century there have been various attempts at dealing with these issues. The behaviorist, for example, declared the inner life basically null. Scientists, on the other hand, did not want to deal with consciousness or subjectivity at all. Midgley finds such attempts unsatisfactory, to say the least.

Midgley argues that the so-called mind-body problem is at heart not a problem about science or philosophy or theology *per se*, rather it is about "how to think." It would pay to start thinking differently about the dichotomy. We should start with consciousness as a complex phenomenon, for which we need all the help that we can get, from the natural and social sciences, philosophy, and religion as well. But abstractions—like that of abstracting mind from body—will not do. When we talk about consciousness, our concern is the whole person, not just a part.

Even Descartes thought that soul and body had to be linked in some way. He thought that they were so intimately joined, that there had to be a mechanism (as yet unknown) to make this union possible. Still, for him, their separation could not in the last analysis be bridged or salvaged. There was a real distinction, and it was essential, between mind and body, and the challenge to find and explain their integration was almost insurmountable. Descartes had to target two different objectives.

> The first is the separation of the human mind and body, the second their integration into a human being. The demands—separation yet integration—seem to make the project impossible... We might be so successful at separating mind and body that when we put them together again to form a single man (sic), we no longer get a real, natural unity—a full-blooded human subject—but what Descartes calls a mere "unity of composition, and artificial com-

pound... To amend things, we might start by carefully protecting the primality of the full human being.[26]

The problem is not just that there are two substances, very different in kind, but moreover that one (soul, intellect) is superior to the other (matter). The inner or pure substance is the only one to survive death and, therefore, it is clear that it is the better one. Thought is powerful indeed—reason is what it is—*cogito ergo sum*—"I *think* therefore I am."

With this view of things, as Midgley suggests, mind and body started to look more and more like "ship (body) and pilot (mind)."[27] However, to conceive of the body as merely mechanical, a tool or a means for the mind to exist and survive is only part of the story. Eventually, people tried to get rid of the "pilot," that is, the independent and controlling mind, thinking it unnecessary since the "ship" is completely mechanical and conceivably fully automated. Even more, the body as machine could probably be reset at our convenience. Thus, from the doctrine of "rationalistic dualism" that neatly divides the realm of mind or soul (thinking substance) from the realm of body (extension), we ended up moving to "materialistic monism," the belief that nothing exists except matter, albeit matter that is fortuitously endowed with energy and eventually with life.

For Midgley, part of the problem with the above line of thought is that it treats mind or the inner person as a single unit, simple, unchanging, and independent, and not as the complex reality that it is *together* with the body. It is true that we are persons, but we are moving, complex personalities as well, beings that are shaped by our relations with others, the environment, culture and, of course, biology.

---

[26] Joseph Almog, *What Am I? Descartes and the Mind-Body Problem* (Oxford: Oxford UP, 2002), xvii–xviii.

[27] Midgley, "Souls, Minds, Bodies, and Planets," 35. Note, however, that Descartes himself did consider this analogy and specifically rejected it as unsatisfactory (Meditation 6, #13).

The Search for Human Nature

Debates on human nature, first about its very existence, and then about its possible definitions, elements, attributes, etc., are as vivid today as at any other time in intellectual history, and maybe even more so. The way that human knowledge has branched out since the eighteenth century, has added not just lots of new information but also multiple levels of interpretation. One outstanding result has been the flourishing of many new academic disciplines and research fields.

The study of humanity as a subject of inquiry has become complex. It is also very competitive, not least because of the many voices involved. From paleontology to biology, psychology to neurology, philosophy to anthropology, there are a myriad of ideas and opinions about the meaning of being human. It is obvious to me that an interdisciplinary approach, the comparison of notes among disciplines, is the best way to proceed if we want to get somewhere. The anthropological issue, that is, the question about and the quest for a "human nature," is taking us into new and unexpected places. A survey of recent literature on this topic can be roughly divided in the following way.

First, there are those who deny an "essence" to the human, not to say "humanity" in the general. We are said to be the product or by-product of, on the one hand, a series of accidental, non-directional (or not designed) blind processes. On the other hand, we are both the receptacles and channels of a myriad of historical processes that shape in inconsistent or even in contradictory ways a human "nature".

However, not everybody agrees on even the need for this effort to locate or define a human nature. There is a kind of post-humanism that even denies that such a thing as "human nature" exists in reality. I think now of British historian of ideas, John Gray, who in a challenging and at times ranting book[28] proclaimed the crisis of traditional humanism. According to Gray, because of its adherence to the (now discredited) creed of human goodness and progress, humanism, as we have known it since the Renaissance, is basically bankrupt. For him, humanists deceive themselves in thinking the human is still central either to nature or to something called history. Gray believes that on both accounts they are wrong. Part of his argument relies on the new awareness of our closeness to other animal species. But also it seems to him that humanism, which he thinks is basically a Christian creation, believes in the uniqueness and special dignity of the human. Since it owes much to Western Christianity, when the latter has lost its relevance, the former has been reformulated in secular ways but without changing its most traditional assumption: the centrality of the human. Science has proven these ideas wrong, not to say "delusional."

Second, there are those who still search for what may be common to the human experience. Take, for example, the idea of the "quest," our endless asking for what, how, why and what-for of all that is. In brief, we are (always) thinking, questioning beings.[29] However, there is no actual end or goal to the quest, since it is the very "questioning," as some believe, that actually drives and defines us.

---

[28] John Gray, *Straw Dogs: Thoughts on Humans and Other Animals* (London: Granta, 2002). Opinion seems to be divided between those who find Gray's books timely and on target in his criticism of humanistic positivism (putting one's faith in human goodness and its predicament), and those who detect in his inconsequential ranting a mixed bag of rather unconnected topics. I actually believe that he is pointing to something worth considering; but his writing style, almost informal or casual, though entertaining, fails to ultimately convince.

[29] The best example of this perspective that comes to mind is Charles Pasternak, *Quest: The Essence of Humanity* (West Sussex, UK: Wiley, 2003); Pasternak is a chemist and professor of biochemistry.

Third, others assert a strong biological basis for almost everything we are, do, and even believe or hope for. Whether in the shape of some kind of "genetic determinism", or through the multifold evolutionary process that shape human adaptability to life conditions, an increasing number of thinkers speak of the inevitability of the formation of human "nature(s)".

Finally, and for our purpose here, there are changing and welcome notions about what it means to be human and animal. Midgley, for one, says that we humans are ethical primates. She sees the roots of our ethics in our naturally developed sociality, which comes to us from our primitive ancestors. The origins of this sociality could be seen as the outcome of the earliest responses to conflict among primitive human gatherings, a conflict-ridden motivation that emerged from evolution,[30] not unlike what happened in other primate groups. The distinctions between what in the person counts as strictly human and what counts as animal, though commonly taken for granted, is not so easy to determine. For the most part, it seems that only matters of degree separate us from other animals in everything, including intelligence, habits, adaptability, social relations and, not least, genetics.[31]

---

[30] Mary Midgley, *The Ethical Primate: Humans, Freedom and Morality* (London: Routledge, 1996), 172–173, 178.

[31] Here comes to mind especially the book by the world-renowned primatologist Frans de Waal, *The Ape and the Sushi Master: Cultural Reflections of a Primatologist* (New York: Basic Books, 2001). De Waal analyzes those recorded instances in which apes have demonstrated that they have evolved the basics of culture: situations in which apes have been able to use creatively problem-solving solutions, and then have passed on to subsequent generations their newly acquired abilities and traditions, even if in a rather simple manner.

Midgley proposes a kind of "humble humanism"[32] as opposed to a rampant "anthropocentrism." She sees in both traditional philosophy and modern science a kind of blindness toward the earth and all other creatures. The living world is not our exclusive domain as humans. We are not above the natural order of things. We are part of nature. We belong in nature together with every other living organism. From this awareness we could elaborate a solid, pluralistic, and working understanding of human nature and human relationships. What we need is a unified conception of what is rational and what is animal in us. But, to do that, we need to do a critique of reductive views of what counts as animal nature as well as what counts as rational.[33] Especially since the Enlightenment, "animal" and "rational" have been understood as completely opposite poles. For the task of formulating a balanced view of ourselves, we need all the help that we can get: science as well as philosophy and religion are all needed.

What made us human in the first place? Was it the growth of big brains, or the development of clever hands? Maybe upright bi-pedalism (locomotion in two legs),[34] which has definitely played a major role in human development, when it was selected by evolution? Moreover, is any physical development a complete explanation, in any sense? Or is technology or culture what has made the real difference? According to Midgley, there is not one answer, nor do we need only one. She sees more problems than solutions with sweeping theories of human nature: they tend to be

---

[32] This is my own expression and not Midgley's or anyone else's.

[33] By Midgley's own account, this is the main aim of her seminal book *Beast and Man: The Roots of Human Nature* (revised ed.; London: Routledge, 1995), 243–244.

[34] These are the kind of questions that Ian Tattersal, *The Monkey in the Mirror: Essays on the Science of What Makes Us Human* (San Diego: Harcourt, 2002) raises as a guide to his own study and analysis of the topic at hand.

one-dimensional. For instance, whenever someone insists that human nature is "basically sexual, basically selfish or acquisitive, basically evil, or basically good,"[35] the complexity is reduced to any single explanatory principle. For Midgley, this kind of reduction is akin to simplemindedness, which is never appropriate to deal with the richness of the human experience.

Midgley proposes that a sense of wholeness, which is the conviction that the whole of the experience cannot be reduced or understood by isolating any single one of the parts that form the totality, needs to be kept in mind among the many possible candidates for explanation of what defines the human.[36] A sense of wholeness is indispensable, one that goes beyond putting all the pieces together. Yet we cannot do without those pieces of evidence and experience.

Midgley is adamant about saying that we still need some sort of notion of human nature. It is within our reach; the temptation for many theorists is to take shortcuts and therefore end up with a narrow concept. Nevertheless, we certainly know some things about ourselves. We have both the individual and the collective experience. Internal factors like innate tendencies, genetic or otherwise, and not only external contingencies (cultural or environmental) have to be kept in mind and investigated. How did behavioral patterns develop for the human race as a whole? For this inquiry, nature and culture cannot be considered opposites, as has happened so many times. None can be excluded. Besides, what we know about human nature also tells us facts, often crucial, about the range of possibilities open to us.[37]

Despite the difficulties and contradictions inherent to it, we have use for a conception of what means to be human. It

---

[35] Midgley, *Beast and Man*, 55.
[36] Midgley, 15.
[37] Midgley, 73.

is such a conception that allows us to judge ideas about human rights, what counts as humane treatment (of self and others), what morality is, etc. By it we come to better understand our moral reactions as well as those of others; it allows us to respond to suffering. Here a notion of human kinship plays an important role.

Midgley has no doubt that evolutionary theory has contributed much to the understanding of ourselves as biological beings who belong to this earth, who are at home here in this planet and place. We are part of nature, neither above it nor against it. Some understanding of our selves as an animal species will help in this regard, especially when it comes to grasp some deep-seated instincts and motivations that drive our behavior.[38] We are of the earth. We are not fit to live anywhere else. The crucial point for Midgley is that "life is not an accident or an alien invader but something which has grown out of the earth itself."[39] As the current environmental crisis has shown, it makes no sense to "conquer" our home, even less to destroy it. We have continuity with every other living organism, especially animals. This is one of the reasons why the understanding of evolution should give us pause, according to Midgley. By this she means a humble sense of our selves, of our limits as well as our capacities. It is unfortunate that many seem to get a curious over-confidence about our possibilities of taking charge of nature and its order, by manipulating it to our purposes — or the purpose of a few, in any case.

The Importance of Thinking Gaia

Among contemporary developments in evolutionary biology, Midgley finds some that take seriously the conservation of our place of origin and the proper sustenance of

---

[38] Midgley, 158.
[39] Mary Midgley, *Science and Poetry* (London: Routledge, 2002), 204.

every living creature. In Midgley's view, one prominent instance and one of the best candidates for bringing these concerns together is the theory behind "Gaia."[40] As a scientific theory, Gaia is the idea of life on earth as a self-sustaining natural system. First proposed by biochemist James Lovelock,[41] and further explored by other scientists[42] — biologists and environmentalists prominent among them — Gaia thinking has long since made its way into the work of philosophers, theologians, and animal advocates of different kinds. Lovelock is an independent researcher who, in the late 1960s was working for NASA and had been commissioned by the agency to study the possibility of human survival in other planets. Lovelock wondered what it would take to create the proper environment for humans to survive in other planetary systems, by recreating the minimally necessary living conditions. One result of such study was that Lovelock developed his concept of the whole earth as a kind of organism and us as an intrinsic part of it, living in a symbiotic relationship with our host.[43]

Although suspected by some scientists as being more than just science (since to speak of the whole earth as a living organism borders, for many, on the mystical), Gaia theory is nevertheless primarily a scientific description of

---

[40] Midgley, 16–17.

[41] See his seminal work, James Lovelock, *Gaia: A New Look at Natural History* (Oxford: Oxford UP, 1979), which originally summarized his work of more than a decade. The term "Gaia" comes from the Greek word for the Goddess Earth, and was proposed to Lovelock by his friend (the English writer) William Golding, after the former explained his thinking and concept to the latter.

[42] For example, Lynn Margulis, *Symbiotic Planet: A New Look at Evolution* (New York: Basic Books, 1998), 113–128.

[43] For more on Lovelock and the development of the Gaia theory, see Jon Turney, *Lovelock and Gaia: Signs of Life* (New York: Columbia UP, 2004), especially 1–46.

the fine and delicate balance created between the planet and all living things. The activity of living organisms (humans included) keeps regenerating what is needed for survival. Everything from the amount of oxygen in the atmosphere, to the distribution of other gases, to the generation and degradation of nutrients and the many chemical reactions involved, they all interact in a mutually dependable manner. The earth is more than our host; it has produced us. But, by the same token, we are not mere parasites. We actively affect—for the most part for good but it could eventually turn against us too—the conditions for the well being of all creatures.

As pointed out before, Gaia thinking was received with certain enthusiasm by many, but also suspected of mystical or mythical leanings (and therefore not scientific) by others. Although not argued with the same intensity any longer, Gaia thinking has not really disappeared. It has found in Midgley, among others, a strong supporter and one who still sees the possibilities that Gaia thinking presents for an in-depth dialogue that includes scientific, ethical, philosophical, and religious concerns.[44] The thing with Gaia is that it provides a vision, and a metaphor, which serves to correct the strong individualism that has invaded science as well—as with talk of "selfish genes",[45] for example.

Obviously the idea of Gaia is a myth, a symbol. But then so is the sociobiological idea of the Selfish Gene. One of

---

[44] For example, her published lecture on *Gaia: The Next Big Idea* (London: Demos, 2001), but also her comments on this respect in interviews and other media, like that with Liz Else for *The New Scientist* 2315 (03 November 2001).

[45] The seminal work that made the concept of "selfish genes" publicly known is that of Richard Dawkins, *The Selfish Gene* (original ed; Oxford: Oxford UP, 1976). Subsequently, Dawkins' selfish-gene imaginary has been quite abundant in popular publications on genetics, biology, and especially, evolutionary psychology.

these myths emphasizes our separateness from the world around us. The other emphasizes our profound dependence on it. Since wholes are quite as real as parts, there is no reason in principle why we should have to prefer the first emphasis to the second. The choice between them depends on their relevance to our situation. And given that current situation, there seems to me to be little doubt about which of them we most need to guide our thinking today.[46]

According to Midgley, by providing a more accurate view of the earth, it "can give us a more realistic view of ourselves as its inhabitants."[47] Above all, it is a corrective to the modern conception of the earth (provided by the scientific revolution) as basically a big machine[48] that can (and ought to) be controlled and used exclusively for our own benefit.

One of the clear restrictions of an "atomistic" science[49] seems to be in making it hard for scientists to have a larger vision, one where their science plays a role without being a discourse about the whole. To think of science as the ultimate explanation of everything, is actually very reductive, and quite unnecessary. It goes well beyond what science can honestly provides us with. For Midgley, Lovelock's Gaia, as a hypothesis, contains scientific truth as well as

---

[46] Mary Midgley, *Science and Poetry* (London: Routledge, 2002), 17.

[47] Midgley, *Gaia*, 13.

[48] For Midgley, we need Gaia thinking together with modern physics—which speaks of forces and fields and web connections—in order to replace the "Newtonian [mechanistic] world-picture," so much dependent on clockwork images and inert visions of matter; see her latest appeal to consider Gaia in her edited volume *Earthly Realism*, 4.

[49] Midgley defines "atomism" as the "notion that the only way to understand anything is to break it into its ultimate smallest parts and to conceive these as making up something comparable to a machine" in her book *Science and Poetry* (London: Routledge, 2002), 2.

religious potential.[50] But it does not do it by trying to explain it all. Rather it gives a good sense of why all forms of human knowledge are needed in order to paint as complete a picture as is humanly possible. Gaia is not a simple or single explanation, because Earth is a very complex system. Gaia speaks of mutual dependence and cooperation, with organisms able to adapt and improve their circumstances by interacting with others. However, it also speaks of certain fragility. Life is precious, and a real thing. The Earth is unique and worth conserving. It takes first a plurality of sciences and then also of other disciplines, including religion, in order to understand and value such a living system as is our own world. Midgley does not deny that Gaia has spiritual or religious connotations.[51] This is in fact part of its beauty and usefulness. She

---

[50] As referred to by Andrew Brown in his interview with Midgley for the British paper *The Guardian*, and quoted before.

[51] Midgley has often referred either to religion or to religious perspectives in many of her writings. Her many statements in this regard have prompted the question about what is Midgley's religion or religious commitment. Having grown up an Anglican Christian, this is the experience that she knows best. However, she has also acknowledged having studied Buddhist concepts, especially through the mediation of one of her sons, David Midgley. She professes to be more of an agnostic concerning the question about the possibility of belief in, or relating to, a personal God. She does not deny the possibility of the existence of God as such. Moreover, in regard to her understanding of what religion is and of its multiple manifestations, she has confirmed herself to be basically in agreement with the American philosopher William James in his work *The Varieties of the Religious Experience* (reprinted ed.; New York: Barnes and Noble, 2004), especially with James' conclusion in that volume. In an email dated August 17, 2005 Midgley shared with me that she does not think religion is one single clear category. Most important, she agreed with James in thinking that religion is an essential element in human life. James actually thought that it was essential in order to understand human nature.

sees it as a point of contact for a renewed dialogue between the religions and the natural sciences. Besides, the idea that a scientific theory may have other possible uses beyond that of providing so-called "neutral facts" about the world is not new. This idea has been part of at least some aspects of evolutionary theory and its offspring from the beginning of its modern (including pre-Darwinian) coinage.

Science and Religion in Conflict

Midgley's concern is primarily with those doctrines that are believed to be thoroughly scientific but that are not. They use the label of science and cloud of science for their own extremist agenda, meaning, that of attacking and even becoming a substitute for religion.

One such example is provided by sociobiology, especially in its Wilsonian inception. Midgley declares that sociobiology, like Gaia, has a religious angle. Harvard entomologist and popular science writer Edward Wilson (b. 1929) has been open about this point, and seen his brand of sociobiology (which he helped to name and define in the first place) as a unifying field. It purposely brings together other disciplines and, therefore, diverse explanations about life in this world, based firmly on evolutionary science, and in actual competition — by being a more comprehensive and rational myth — with religion.[52] For example, Wilson's call to,

> Make no mistake about the power of scientific materialism. It presents the human mind with an alternative mythology that until now has always, point for point in zones of conflict, defeated traditional religion. Its narrative form is the epic: the evolution of the universe from the big bang of fifteen billion years ago... to the beginnings of life on earth....
> As I have tried to show, sociobiology can account for the

---

[52] See Midgley's insights on this regard in *Science and Poetry*, 199–200.

very origin of mythology by the principle of natural selection acting on the genetically evolving material structure of the human brain.[53]

For Wilson, the unifying myth—not to be confused with a lie, but rather thought of as an imaginative vision—is his idea of a "scientific materialism." The latter has the power to present the human mind with an alternative mythology, a consistently meaning-making narrative, guided by the corrective nature of the scientific method, and one that has finally defeated religion. According to Midgley, this is undeniably faith.[54] Why? Because it goes beyond itself to find comfort in the promise of an ever new and better present and a reliable future. Midgley notes,

> It is a faith to suit the states of its congregation. This consideration is evidently strong, since it is hard to see what scientific grounds Wilson could possibly offer for expecting the future to be better than the past. That expectation did, of course, figure in Lamarck's and Herbert Spencer's view of evolution. But scientists are supposed now to have abandoned it. Progress of that kind forms no part of Darwin's doctrine and current science says nothing to support it.[55]

Midgley's contention is that scientific mythology, if recognized as such, may have a place in public discourse, but it will always be a partial vision, not a final universal truth. Even among different scientific disciplines—say, physics, biology, and chemistry—there seems to be competition for a priority in explaining the world in general and life in

---

[53] Edward O. Wilson, *On Human Nature* (Cambridge: Harvard UP, 1978), 192.

[54] For Midgley, faith is not merely a collection of beliefs; it is a way of relating oneself to a bigger or greater whole, as being part of it, and somewhat dependent on it. This faith is not itself a religion but the "seedbed of religion." See, Midgley, *Evolution as a Religion*, 17.

[55] Midgley, *Gaia*, 34.

particular. Each discipline believes its work is more fundamental than the other. Sometimes it is understandable that this anxiety of competition spreads out to other fields or views.

For Midgley, however, to actually believe that science and religion are bound to a kind of deadly winner-takes-all match is misleading. As a matter of fact, science and religion represent, each in its various manifestations, different levels of reading the world of our reality.[56] Moreover, they are not things or objects that we find out there in the world, but fields of enquiry, modes of interpretation, full of difference and nuance.[57] In nature, contrary to some kinds of old biological thinking, cooperation is as abundant, and in some instances even more present, than competition. It is only reasonable to think that it can be so in human affairs.

Science and Religion Matter

Science seems to have different meanings. It is sometimes taken as a code word for "rationality," and at others for "factuality." Or it could be constructed negatively as "non-superstition" or as "non-faith." For many, it is just a vast memory store, or a register of facts about the world. So it also happens with the term "scientific." Midgley explains that it could be construed as a term of praise, as meaning "thorough," "systematic," or "methodical." But it could also mean simply "concerned with the natural sciences" as opposed to other studies."[58]

Generally speaking, science is a discipline as well as an activity. As a discipline, it follows some established meth-

---

[56] On this issue of different levels of reading the world, as well as a proposal for a hybrid reading of reality, see John Haught, *Deeper than Darwin: The Prospect for Religion in the Age of Evolution* (Boulder, Colo.: Westview, 2003), especially chapter two, 13-25.

[57] Acknowledgement is here given to Dr. Khalid Blankinship for what I consider a "good turn" of expressing it.

[58] Midgley, *Science and Poetry*, 144.

ods or conventions. As an activity, it is normally regulated by a number of institutions, associations, or societies. The Ancient Greeks and Romans had terms for science-like activities: *episteme* and *scientia* respectively. For centuries in the Western tradition the study of the natural world as well as theories about reality were held together under "natural philosophy."[59] The latter term was in use for many centuries, well into Darwin's time. However, since the end of the nineteenth century, science has been thought of, in addition to the previous meanings, as a worldview. Therefore, to talk about a "scientific outlook" on the world has become increasingly common in many societies, and not just in the West.

For Midgley, science is basically an intellectual system constructed by reasoning with the aim of understanding the universe, the world of reality in its individual parts. One of the major contributions of this "scientific rationalism" is in helping us to avoid treating the world of experience as something unreal or delusive in any way.[60] There is a real world out there. Science builds knowledge, and by so doing, constructs a meaningful experience of the physical world for us. And it does it in a twofold manner.

This brings up a matter of the greatest importance about the nature of science itself. *Science always oscillates between*

---

[59] For more on the meaning(s) of science, its history and applications, see the introductory treatment by philosopher of science Del Ratzsch, *Science and Its Limits: The Natural Sciences in Christian Perspective* (Downers Grove, Ill.: Inter-Varsity Press, 2000), especially the arguments in the first six chapters, 11-99.

[60] Midgley, *The Myths We Live By*, 27-28. She makes this point with force against the "extreme constructionists" — those who consider science to be an elaborate discourse, a constructed view of the world, but another product of specific social circumstances, owing basically to its political context its tendency to turn points of view and their ensuing metaphors into statements of fact about reality — and their bland conception of science.

two magnets, two equally important ideals. On the one hand it aims to represent the hugely complex facts of the world. On the other, it aims at clarity, and for that it needs formal simplicity. When mathematicians are in charge, the second ideal always tends to dominate over the first. And, for a long time, mathematics provided the only model of intelligibility that physical scientists saw how to work with.[61]

To this day the tendency is to think first of physics as the basic definition of what science is. This dominance of physical science is partly due to its close relation to mathematics and the idea that it is "exact" or "hard" science, that is, almost flawless in its predictions about the world.

In the case of evolutionary biology, however, beginning with Darwin himself, and despite the attempt at simple principles of explanation, there has been awareness that the complexity of life forms, the relations between organisms and their environments, and the historical character of the science itself, do not hide the challenges and difficulties inherent to their full understanding. Science needs more than deductive reasoning and empirical evidence to do its work. Inference, intuition and not a little dose of imagination are needed as well.[62]

Without falling into any kind of post-modern relativism, Midgley asserts that some faith is unavoidable in doing science. The scientist is compelled to trust the world as it presents itself to be discovered, analyzed, and explained. Faith is applied to choices, on how we regard the universe, and the order of everyday experience, for example. However, science would not be able to proceed with uncritical *belief.* This much is obvious. Science always has an eye to test

---

[61] Midgley, 129.

[62] No scientist has demonstrated this point more forcefully than Einstein and the Quantum physicists themselves. It tends to change the very nature of evidence in science. It also brings to mind Einstein's maxim that "imagination is more important than knowledge."

its conclusions. By the same token, unexamined *unbelief* is also damaging to science. First, because questions that can have no answers become fraudulent and without meaning. Second, and most important, because doubt is then applied selectively to cover only those areas in which scientists are either reluctant or don't see the need to deal with. This latter problem, as we will see in more depth later on, Midgley thinks to find in Darwin himself.

Thus, we can safely say that science asks questions, conveys information, and therefore, it offers not just to understand the world but to change it too.[63] It is also true that a measure of "methodological skepticism" is always proper. Although it has often been repeated in many circles, the idea that science deals strictly with facts and religion (or ethics) strictly with values cannot be justified. We humans do not necessarily work that way. Both science and religion deal with facts, and they both help us to react to the world as it presents itself in all its complexity. If we still think in the traditional way, it may be partly due to the old assumption that science is all about reason, and that religion is about feelings or emotions. Furthermore, we would also be assuming that reason and feelings are two very different things and therefore ought to be kept separated.[64]

But this is not to be so, according to Midgley. Why? Because multiplicity of motives, as well as multiplicity of aims, can be very rational indeed, especially when we are

---

[63] With a reference to Marx (and against Marx), Midgley is actually correcting the idea that we have to make a choice between practice and understanding—there is no need to split our experience in this way.

[64] See Mary Midgley, *Heart and Mind: The Varieties of Moral Experience* (revised ed.; London: Routledge, 2003), 5–8. Midgley sees in various philosophies, for example, in that of David Hume, the assumption of the separation between feeling and reason, and the supposed need to choose between one and the other in many areas of life.

made aware of possible and sometimes inevitable conflicts. To insist on separating facts from values, or the objective from the subjective experience, or reason and feelings, ends up splitting our world and enhancing conflict, instead of coming to terms with it. We will force ourselves then to choose one side or the other.[65] By having to choose this way, we will never be whole, even though wholeness is indispensable to the life of our species. A sense of wholeness as well as cognitive integrity cannot be dismissed as illusions. But splitting the self will give us the false hope of simple explanations for complex matters.

However, the fact remains that simple premises can distort understanding and lead to greater confusion or even irrationality. And not even the practice of normal science, not even if we hold onto the conception of a "pure science" for that matter, has been able to escape this fate.

Can Science and Religion Co-Exist?

To think clearly about the relationship between science and religion matters to Midgley. But beyond her concern, it is an important question and discussion for our time. There is a lot at stake in sorting out two of the most important human endeavors that have a strong impact on today's world. At face value, science and religion seem to be two very different activities and forms of thinking. However, we could safely assume that there have to be some questions that are common to both experiences. After all it is one and the same human person who is asking the questions. As we just saw above, the idea that science deals with facts and religion with values, as many seem to assume, does not actually help in keeping together the whole of the human experience. If anything, we could at least think of both disciplines as following tracks of enquiry, probably even similar pursuits. Is a constructive dialogue possible, or is conflict between the

---

[65] Midgley, 146-147.

two inevitable? Does the advancement of science imply the retreat of religion? These are but some of the questions that are raised by comparing the two.

I would venture to say upfront that as long as there is wonder, and questioning beings like us, both religion and science are here to stay. Moreover, that it is science as a "belief system" that gets into conflict with religion. Some would argue that tensions do not necessarily disappear, even if we hold an enlightened kind of religious belief.

As a way to start "thinking clearly" on this topic, let me begin by discussing four possible forms that the relationship between science and religion may take. For this, I want to refer initially and briefly to what physicist and theologian Ian Barbour (b. 1923) has proposed in several of his writings.[66] Then, we go back to what Midgley has to say.

First, the relationship between science and religion can be and has been viewed in terms of "conflict." Especially since the late nineteenth century, the argument has been made for a so-called "war" between religion and science.[67] However, this view seems to have its deeper roots in the Enlightenment of the eighteenth century. This perspective of inevitable conflict conveys a rereading of the history of both fields in which religion has represented the forces of obscurantism, and science the forces of rationality. Among the chapters of history that have been revised in this "light" is the

---

[66] See especially Ian Barbour, *Religion and Science: Historical and Contemporary Issues* (revised and expanded ed. of *Religion in an Age of Science*, 1990; New York: Harper San Francisco, 1997), especially chapter four, 77–105. Although many other books have been written in the last two decades or so, which provide a number of useful ways to understand this relationship, I believe that Barbour's is still a reliable way to argue about it.

[67] The two best examples are John William Draper, *History of the Conflict between Religion and Science* (1874), and Andrew Dickson White, *History of the Warfare of Science with Theology in Christendom* (1896).

story of Galileo, portrayed as a champion of reason against religious superstition and ecclesiastical authoritarianism. The latter case is the belief that this "war" escalates when religious authorities impose a certain, and unproven, "scientific view" of the world.

This is not the place to enter into the details of the story, but it will suffice to say that the Church was actually the primary sponsor of science, and especially of mathematically oriented science, physics and cosmology, in the late Middle Ages and into the early modern period.[68] The Galileo controversy was as much a theoretical argument regarding true certainty and the nature of proof as it was a conflict of personalities. But it was not necessarily a "war" between science and religion.

Second, the relationship between science and religion can be viewed in terms of "independence," meaning, these are different disciplines, with different objects of study, and they use their own methodologies. They both have their own language, which is used to describe reality. In this sense, no real conflict is possible, or any comparison. Science is understood as a cognitive function, the understanding of the world. Here science is about knowledge and truth. Religion, on the other hand, is defined in terms of non-cognitive functions: it explores the emotional, psychological side of human experience. This second perspective has many adherents. At least it does recognize that the human experience is multifaceted, and that different languages (in the conceptual sense) are needed to describe the elusiveness of "reality."

The third way to describe the relationship is through "dialogue." Here a certain basic independence is recognized for each field, and none is subordinate to the other. It

---

[68] On this point, I refer the reader to the persuasive arguments made by Margaret Wertheim, *Pythagoras' Trousers: God, Physics, and the Gender Wars* (New York: Norton, 1997), a fascinating and beautifully written book.

also recognizes that there are certain areas of human experience and thought in which the two disciplines overlap. Parallels between methodologies are studied and emphasized. It allows for a relation of enrichment between the two, even better, for the recognition of a possible interdependence in some issues.

Religion points out to questions of transcendence, beyond common sense-experience; it relies more on intuition than science, but then, it makes clear how much science itself may depend on intuition in the elaboration of hypotheses. By the same token, science shows religion the importance of a systematic method to survey and interpret experience,[69] which is basic data also for religion. The need for dialogue is especially strong when considering the fact that theologians do make assumptions about the universe and that in many instances these assumptions are based on old cosmologies. Dialogue allows at least an updated picture of the world. Otherwise, theology risks becoming irrelevant intellectually speaking.[70]

---

[69] In my opinion, one really good example of applying a method derived from science, is that of the German scholar Gerd Theissen, *Biblical Faith: An Evolutionary Approach* (trans. John Bowden; Philadelphia: Fortress, 1985), where he makes use of Darwinian evolutionary mechanisms, such as selection, variation and adaptation to account for the development of certain biblical concepts. We could argue about the validity, not to say the necessity, of such an approach, but I certainly think that Theissen has done a good job in the application of these mechanisms in his analysis of some New Testament teachings.

[70] See also John Haught, *Science and Religion: From Conflict to Conversation* (New York: Paulist, 1995), especially 17–21. Haught collapses under one label, "Contact", Barbour's third and fourth way, by insisting that he does not see a real different between the two. More than openness to the other, Haught argues for a position of seeking contact between science and religion that recognizes the distinctiveness of each approach, like, for example, in some aspects related to method and language. In the

The fourth option is that of "integration."[71] Though the option of "dialogue" seems to be very popular today, an increasing group is claiming for a relation of mutual dependence between science and religion. There is recognition that the methodologies are not that different after all, even though they have different objects of inquiry. This view points to the unity of the human experience of the world. Interpretations vary, since context and historical conditions directly affect them. But this is the same for both disciplines. Science and religion are both elements of the totality of the human experience of the world. Although Barbour has been persuasive in making the case for serious dialogue and exchange between science and religion, not everybody agrees that integration can be a serious call.[72]

With this trust in mind, religion seems to be closer to the "epistemological roots"[73] of scientific inquiry than is the

---

real world, in actual life experience, these two forms of inquiry have influenced each other in many ways.

[71] Haught's fourth concept is "Confirmation", where he argues for his own conviction, that of discovering consonance between science and religion as the most important work that needs to be done; see Haught, 21–25.

[72] On this point John Haught argues that religion can confirm science in its own pursuit for knowledge, the "desire to know reality," as well as in its basic trust on the rationality and intelligibility of the world. Religion then provides confirmation to that trust. That confirmation does not imply that religion provides science with facts or that it can do the same job of exploring nature that science knows how to do and can do on its own. See Haught, 21–23.

[73] This is one point in which I agree with Haught and if I take the opportunity at this point to mention Haught's argument here, it is because it will be treated more extensively in the context of later chapters. There it will be argued that some of the problems that religious people, especially Christian literalists, confront with evolution, is intimately related to an epistemological issue:

case in the other approaches mentioned before. Religion does not create the trust, since that trust is part of our own natural cognitive function. We seem to have a natural disposition to relate to everything else in the world. However, religion encourages, keeps alive, and renews such trust whenever necessary to do so. Based on experience, we can say that religious faith is a kind of re-assurance, a reinstated confidence in living despite all of the negative factors and tragedies of life. Nevertheless, Barbour's basic argument is that both science and religion can at times share similar methodologies. They could both answer to similar questions or concern.

In order to make his case, Barbour refers to the use of concepts such as "model" and "paradigm" as ways in which both science and religion make themselves understood to others.[74] Both science and religion use models in their own hypothetical and theoretical constructions. These models are conceptual representations that convey an idea or ideas about a subject matter. A model is an analogical device. Because reality cannot be described in direct terms, and because any experience of the world is always a mediated experience (since all observation is "theory-laden"), then we use metaphorical constructions in order to give a better sense of what is out there. The metaphorical constructions that become permanent allies of an interpretation of the world of experience are called "models."

Models are used both in science and in religion. In neither case have we anything like a direct access to reality. Models are somehow adjusted to the experience (e.g., observation, data), but models can also lead to a different perspective or kind of experience. According to Barbour, models convey meaning. Yes, they are of human construction, theoretical

---

certain assumptions and a certain approach to matters both divine and human.

[74] Barbour, *Religion and* Science, see the section in 106–136.

constructions, but to say so is just to recognize that theory (understanding) cannot be detached from experience. There is no pure experience of the world in that sense. And the same happens to the scientist as well as to the theologian or a religious person.

The other important concept that Barbour uses is that of a "paradigm," taken basically from the philosopher of science Thomas Kuhn (1922–1996). A paradigm is a tradition of scientific research that is passed on through the use of "exemplars", i.e., through concrete examples of how scientific research is done which are received as authoritative by a (scientific) community. In Kuhn's words: "I take [paradigms] to be universally recognized scientific achievements that for a time provide model problems and solutions to a community of practitioners."[75] Paradigms may suffer internal changes, but they still survive generations of researchers. A new paradigm is a substitute for an older one by nothing short of a revolution in thought, e.g., the "Copernican revolution" in astronomy.

The concept of a paradigm is as useful to understanding tradition and change in religion and theology as it is in science. However, there have been quite a few critics who have challenged the adequacy of explaining consensus and change in science through the use of "political changes" and "contextual determinism" which seem to define pretty much what a paradigm is all about. This is not a complete rejection of the concept, but it has to be revised and expanded in view of new understandings in both philosophy of science and philosophy of religion. Among those, the so-called "linguistic turn" in epistemology,[76] and the scientific tendency

---

[75] See Thomas S. Kuhn, *The Structure of Scientific* Revolutions (3rd ed; Chicago: The University of Chicago, 1996), x.

[76] Here I rely on the basic definition given by Christopher Norris, *Epistemology: Key Concepts in Philosophy* (London: Continuum, 2005), as "the idea [coming from Wittgenstein] that truth-claims

toward beauty and simplicity as exemplified in the search for unitary theories or "theories of everything."[77]

According to philosophers of science, Mary Gerhart and Allan Russell,[78] the stride toward beauty and simplicity in science is a sign of the presence of a religious dimension in the work of the scientist. On the other hand, the concern of the theologian to understand religious experience in a systematic and epistemic way is a sign of the influence of scientific concerns with analysis and method in science. These authors want to make the case for a closer look at the interrelation between science and religion. They argue that time has come to move from mere parallels between the two toward a true exchange of ideas, motives, and methods. They argue for a "process epistemology" or "knowing-in-process" that can be applied to both fields.

As also happens in Barbour, Gerhart and Russell argue that science and religion are expressed through metaphors, using models, and being situated within specific paradigms or research traditions. However, they have a more

---

of whatever kind... are all bound up with our manifold 'language-games', cultural practices, or 'forms of life' and are therefore to be judged each by its own *sui generis* criteria of valid or meaningful utterance"; 6.

[77] See, for example, a basic and very readable history of the pursuit for unification in physics and cosmology by Dan Falk, *The Universe on a T-Shirt: The Quest for the Theory of Everything* (New York: Arcade, 2005).

[78] See Mary Gerhart and Allan Russell, *Metaphoric Process: The Creation of Scientific and Religious Understanding* (Fort Worth: Texas Christian University, 1984). The authors have expanded their analysis to the discussion (important also to Midgley, by the way) of the use of maps in attaining knowledge across disciplinary boundaries; see *New Maps for Old: Explorations in Science and Religion* (New York: Continuum, 2001). It is only unfortunate that the examples given in both texts are limited to physics, cosmology, and mathematics, with nothing on biology or evolution.

expanded understanding of metaphor than Barbour's. For Gerhart and Russell, metaphors are not merely analogies. An analogy is a comparison between something known and something unknown for the sake of understanding. A metaphor is a comparison between two unknown and seemingly unrelated things; it is a distortion of the field of meaning, through which new meanings and possibilities are created.

Therefore, knowing-in-process is a "metaphoric process," always adding new perspectives and understandings, in brief, knowledge. These authors emphasize the cognitive dimension of religious experience. They call for science and religion to work together for a richer understanding of the human experience of the world.

The writers that we have thus far mentioned in this section all share the view of "critical realism," even though their emphases are somehow different. They all reject, on the one hand, the naive realism that takes "direct experience" as the foundation of true knowledge. In both science and religion, experience is mediated either by instrumentation ("indirect experience" in science) or by tradition and community (in religion and theology). On the other hand, they equally reject "relativism," either through its representation in the philosophy of social constructionism, or through the overemphasis on linguistic and literary-critical approaches to the discourse of both "fields of meaning."

As philosopher Anthony Flew explains: "Critical realism retains the belief of commonsense realism in independent physical things, but admits that these are not directly and homogenously presented to us in perceptual situations ... In general, critical realists hold that knowledge of the world can be gained because there is some sort of reliable correspondence between sensa, or some sort of intuitive data, on the one hand, and external objects on the other."[79] Critical

---

[79] See, Anthony Flew, *A Dictionary of Philosophy* (revised second ed.; New York: St. Martin's, 1984), 81.

realism then takes seriously the "givenness" of experience — it is not pure imagination — as well as the importance of context and interpretation in the construction of knowledge. Theory always has priority over experience, but at the same time, theory is dependent upon experience and limited by it. In other words, we have both "theory-laden experience" and "observation-inspired theory."

Midgley's Take: No Substitute for Science or Religion

A lot of what passes for the so-called war between science and religion is borne out of Western intelligentsia's reacting against traditional Christian values and power. What surprises Midgley is not this reaction *per se*, but that it seems to be somewhat old. The situation for Christianity in the West and other parts of the world has changed dramatically.[80] Science is already occupying the place and role that religion, especially Christianity, used to play in many societies.

In the past, there were those thinkers who either moved from science to religion, for instance, in the worship of the divine, or who were inspired by their religion to do science. Nowadays, however,

> The emphasis is no longer on knowing things that are themselves of special value, but simply on knowing something securely, exactly, with absolute certainty. Increasingly, too, this value has been treated as self-evident, self-justifying, needing no explanation. Today, accordingly, we quite often find a strange situation where knowledge about the physical universe is treated as obviously the supreme human achievement, although the thing known — the physical universe itself — is regarded as mere dead

---

[80] On this and related issues, see Midgley's essay "Strange Contest: Science versus Religion," *The Gospel in Contemporary Culture*, ed. Hugh Montefiore (London: Mowbray, 1992), 40–42. I will follow some of Midgley's arguments herein with my own comments.

matter, having no value at all, and no creator behind it. What is venerated is then simply the human scientific intellect. Science is thus revered by people who, in theory at least, do not revere anything else, not even the object of science. Its value is displayed, not as part of a wider pattern of human ideals, but in stark isolation.[81]

Science is so much part of the way in which we look at the world *and* talk about it, that its impact on our worldview is sometimes, by the fact that it has become so familiar, overlook. At times, the very idea of science is held up for intellectual enquiry itself.[82]

Present-day culture takes for granted that there is an open contest between science and religion, as if we only need one for the role and functions that each one fulfills. One contributing factor to this view is our emphasis on knowledge over any other value or ideal.[83] A related factor is that a hierarchy among different kinds of knowledge is taken for granted. What the latter implies is that not all knowledge producing activities or disciplines are valued in the same way. For example, we assume that science is more essential to human wellbeing and plainly a better, more reliable activity than, say, literature or religion, or that mathematics is a higher-level function than art or music. Even within the sciences, such an assumption of a hierarchy of intellect or value seems to be in use, posing physics over and sometimes against biology or geology, to name a few.

According to Midgley, another factor contributing to the split-level view happens when knowledge becomes isolated and stops to be considered a value among others and is thought as disconnected from other values or activities. Thus, intellect reigns supreme. Again, together with Des-

---

[81] Midgley, 42.
[82] See also by Mary Midgley, *Wisdom, Information and Wonder: What is Knowledge for?* (reprint ed.; London: Routledge, 1995), 12.
[83] Midgley, "Strange Contest," 43–44.

cartes, and in a twist to his original concern, we still proclaim, "I *think* therefore I am." Moreover, the concept of knowledge has changed. The old assumption that the accumulation of knowledge was but a path towards the attainment of wisdom has been supplanted by the narrower ideal of science.[84] All knowledge was taken to be conducive to wisdom, which was to serve the needs of human life, the whole person, including the inner or spiritual dimension. Now true, reliable knowledge is thought as exact and quantifiable information about the physical world.[85]

The conception of science itself has changed. It has come to be primarily understood as information, facts about the physical world, acquired by experimental methods. If science were constrained only to, on the one hand, the pursuit of knowledge for its own sake (like art) or, on the other hand, to practical applications or consequences, it would be just fine, and there would not be any notion of science competing with, or superior to, anything else. However, science is seen also as an activity with high spiritual value or as an end in itself, which, as a world-view, is brought then into conflict and competition with religion.[86] As a matter of fact, the competition with religion has been very much emphasized due to the peculiar relation between Western societies and Christianity.

Science can be neither the supreme value nor the only value to live by. As Midgley ponders, who can actually live their life treating the quest for scientific facts (or truth) as being above other values, over love, justice, freedom, or art?

---

[84] Midgley, 41.

[85] Granted that, despite of Midgley's argument, this is not the only meaning that information has in scientific parlance. Still, information as accumulated evidence about the physical world remains one of its primary uses. See, for example, Hans Christian von Baeyer, *Information: The New Language of Science* (Cambridge, Mass.: Harvard UP, 2004), 9.

[86] Midgley, "Strange Contest," 41.

Basically, someone who feels this way would have to be a true believer in the life-guiding or life-sustaining value of science. But the fact remains that we can find true believers in religion as well as in science.[87]

Science is not merely a collection of facts but a value. It is a way of looking at things, more like an attitude. It has its place, and fulfills its role, among other important human activities. If we separate fact from value, more often than not we are going to end up treating all facts as scientific facts. For Midgley, the latter is clearly nonsense. There are many more facts about the world and experience, about reality and life, about ourselves, which are not strictly speaking scientific or for which science does not have anything to say. This is especially the case in those areas of the human experience where science had excluded itself for long. Art, religion, and psychology are able to explain facts and convey meaning. These are not necessarily forced to express meaning without facts, or unveil facts without meaning. Likewise with science: its facts are not necessarily deprived or independent of all meaning and value.

Of course, as some people might say, "there is science and there is science," meaning, that not all scientific activities are the same or contribute in the same way to knowledge of world and self. Therefore, there is not one science, as there is not one view or experience that is able to explain it all, or make sense of it all. To think otherwise is an illusion.

Questions about what we are and how we ought to live would now (they say) no longer need religious solutions, no longer involve agonizing personal doubts and struggles. They could be settled vicariously and impersonally by suitable trained experts. This hope could not last. The mere

---

[87] On this and related points, see Chet Raymo, *Skeptics and True Believers: The Exhilarating Connection between Science and Religion* (New York: Walker and Company, 1998). Raymo argues that the connection between science and religion is better described in this way: that in both we find "skeptics" and "true believers."

number of different, conflicting supposedly scientific solutions to life's vaster problems soon undermined all the claims to final, impersonal authority.[88]

Midgley supports a pluralistic vision of science while at the same time understanding the important role that science plays in our societies. Yet, to expect science to do it all for us requires a strong faith in science's power. We would surely be deluding ourselves to believe so.

---

[88] Midgley, "Strange Contest," 45.

Chapter Three

How Come Evolution?

What is Darwinian Evolution?

With the publication of *On the Origin of Species*[1] in 1859, Charles Darwin introduced more than just a new take on evolutionary thinking, which had a history of its own before Darwin. He actually transformed science (or natural philosophy, as it was known then) into a respectable, full-time, public (not merely academic) profession. Others had written about evolution before Darwin, including his own grandfather, the physician and poet Erasmus Darwin, and thinkers of the stature of the French scientist Jean Baptist Lamarck. But Charles Darwin converted evolutionary views into a systematic programme of research.

Darwin's basic idea and formulation are simple enough for the common person to follow. In truth, Darwin has given us more than a theory, but rather a "bundle of different—yet related—ideas."[2] Still, the basic tenets of the Darwinian theory are rather facile to explain.

---

[1] Charles Darwin, *On the Origin of Species by Means of Natural Selection, or the Preservation of Favoured Races in the Struggle for Life* (reprint ed.; New York: The Modern Library, 1998).

[2] According to Richard Morris in *The Evolutionists: The Struggle for Darwin's* Soul (New York: Henry Holt, 2001), 53; on this point,

They begin with the assertion that life on Earth has evolved. Species (plants, animals, or bacteria) are not static organisms. They are rather in states of permanent change. Those changes can be confirmed by the fossil record (historically) as well as by observation (experimentally).[3] The truth is that this assertion about the basic "permutation" or transformation of species did not originate with Darwin.[4]

What follows is the affirmation that life is a struggle. Organisms struggle for their survival, by looking for and securing their sustenance. Moreover, organisms struggle also to reproduce. They all seek to leave the greatest number of offspring. However, at any given time there is more organic life, more individuals of any and all species, than could actually survive. Darwin learned from Thomas

---

Morris follows the lead of famed German American biologist Ernst Mayr.

[3] See, for example, John Dupré, *Darwin's Legacy: What Evolution Means Today* (Oxford: Oxford UP, 2003), 12.

[4] See, Edward J. Larson, *Evolution: The Remarkable History of a Scientific Theory* (New York: The Modern Library, 2004), 13–15, 66. Larson argues that during the eighteenth century, ideas about organic evolution were introduced in the writings of the French scientist Georges-Louis Leclerc, comte de Buffon (1707–1788), especially in his treatise, *Natural History*, published in 44 volumes over an extended period of his productive career. Buffon developed his conception of a materialistic origin to life and species as an alternative to the traditional Christian views of creation. After Buffon, it was Erasmus Darwin (1731–1802), Charles Darwin's grandfather, who took the banner of organic evolution and presented it in a highly original and poetic work, *Zoonomia*, published between 1794 and 1796. The problem with both of the predecessors, though not really precursors, of Charles Darwin was that their work contained very little or no scientific research properly speaking. Above all, it was Jean Baptiste Pierre Antoine de Molet, chevalier de Lamarck (1744–1829) who offered a scientifically argued, although not always credible, "transmutation hypothesis." More about Lamarck's views later in this chapter.

Malthus, in his *Essay on the Principle of Population* (1798), that the competition for food is hard and becoming harder with every generation, since organisms tend to produce more offspring than the environment can actually sustain. There are not enough resources for all. They cannot possibly all survive.

There ensues a struggle for life, where some organisms are better equipped (meaning, those who enjoy advantages within given circumstances), and are therefore better adapted to present conditions than others. According to Darwin, those that are better adapted are "naturally selected" for survival. As has been remarked many times, this is a kind of "tautology," and that is true—it says that those who are fit to survive do so, and those who survive are therefore more suited. But all this shows is that the Darwinian principle to account for evolution is commonsensical and lacks a concept of design or direction, which as a science, it does not really need.

Owing to their very success, organisms pass on those characteristics to their descendants. The fundamental principle at work here is "descent with modification." It means that those advantageous characteristics that are preserved are then passed on to the next generation. It is this principle of preservation that Darwin calls "natural selection."[5] This explanation by natural selection is, properly speaking, Darwin's main contribution to the theory of evolution.

It is important to stress that natural selection, as understood by Darwin, is neither a conscious "force" nor a "creator" in any useful sense of the word. It has normally been conceived of as a blind and mechanical process, seemingly driven by chance. As philosopher of science Daniel Dennett (b. 1942) insists, natural selection is an algorithm, a repetitive process, and an unconscious and natural device, which has proven to be effective in the almost endless generation

---

[5] Darwin, *Origin of Species*, 108.

of new organic forms. The choice of language here is Dennett's own. He thinks of an algorithm as a foolproof recipe, a step-by-step long and tedious process, which always does what it does well, though mindlessly.[6] I for one have found Dennett's analogy helpful; but I can also acknowledge its difficulties. As an applied analogy, it makes a number of assumptions already about what natural selection is and is not.

Another basic tenet of Darwinian theory is to point out that evolution through natural selection requires an incredible amount of time to happen. The process is gradual and constant. In Darwin's time, the science of geology had developed enough for everyone to understand that the Earth was much older than previously thought.[7] Dating the Earth had become almost like a sport: scholars kept pushing the date back time and again. In any case, it was clear to Darwin that only a very old Earth could account for the great variety of organisms, extant and extinct, and therefore of species.

Finally, Darwin believed that the variations (or mutations, as he called them) needed for organisms to better adapt to changing circumstances and therefore to survive and reproduce, were random and not planned. These changes happen rather by chance. Moreover, there are plenty of such variations. Naturally, a process by which

---

[6] Daniel Dennett, *Darwin's Dangerous Idea: Evolution and the Meanings of Life* (New York: Simon and Schuster, 1995), 50–51.

[7] Darwin's greatest debt in this regard was to Charles Lyell (1797–1875), whose three-volume work, *Principles of Geology*, published between 1830 and 1833, was very influential at the time. Darwin took a copy of this work with him on his voyage aboard the *Beagle*. Lyell's geologic 'uniformitarianism' taught that changes in the earth (geological strata, for example) have occurred through a long and gradual process and not through the means of regular catastrophes. This idea was extremely important in the development of Darwin's thinking on evolution. See, Larson, *Evolution*, 46–50.

some variations will be established by their enhanced survival rate will ensue. Therefore, by relying on a purely naturalistic understanding of these processes, Darwin had no need for either supernatural or teleological explanations in his science.

Darwin built up his theory from the facts available to him. We can say that he was a "bottom up thinker."[8] He himself accumulated many specimens, fossil and organic, which he then studied with attention to detail. He did it, first, during his five-year ship trip (in the *Beagle*) through the east coast of South America and up to the Galapagos Islands off the South American west coast. During those years, he kept sending many of the collected specimens home. In addition, even after the *Beagle* trip, he worked as a naturalist through years of intense study of different organisms. He wrote about and became an expert on a number of species: finches, barnacles, and earthworms are probably the best-known cases. Moreover, through an extensive correspondence with farmers, breeders and fellow scientists, he sought information about the behavioral patterns and reproductive habits of many animals in their own adaptive environments.

What matters now is to understand that Darwin considerably advanced the inductive approach of doing science. Before Darwin, deductive methods were of common use. With the latter kind, one proceeds by drawing conclusions from reliable assumptions or facts, either rationalized by logic or proven by experiment. This is the type of deduction used in mathematics, which had shaped much of what was practiced as science since the seventeenth century.

The kind of work that Darwin was doing, by gradually accumulating an overwhelming amount of evidence, and

---

[8] I use this term, as applied to Darwin, to mean someone who moves from a scientific perception of nature (inspired by wonder) to a sense of awe and reverence for nature; second, for someone who builds his ideas from the most basic and simple towards the more complex.

from there starting to build theories (again, from the bottom up), was already acknowledged during Darwin's time as a remarkable and convincing way to reach scientific knowledge. Early on, during his research as a naturalist in the *Beagle*, Darwin looked for gradual development among the myriad organisms that he studied and the thousands of specimens that he collected. This was his "up and coming approach."[9] There is a sense in which Darwin saw everywhere in nature this kind of repetitive, gradual, and consistently simple mechanism. As Dennett says, it is nothing short of an algorithm.

Nevertheless, Darwin made an impact on more than a theory—already known as developmentalism, transmutation, or simply as evolution even before Darwin's time. According to Ernst Mayr, what Darwin did with his *Origin of Species* was to effect the secularization of science.[10] Darwin's impact has extended to the way science is still done for the most part today.

Evidence for Evolution: Only A Theory?

Before the publication of the *Origin of Species* naturalists had long been debating how to better assess the many fossils already found, and the many more turning up, in many places. Because of the Christian culture of the time, any explanation as to the origin and meaning of the fossil evidence had to wrestle, among other things, with the biblical record, which was actually almost non-existent in this regard. Early on in the debate, there were voices insisting that the Bible was not a dependable source to sort out this kind of evidence. Moreover, it was already argued that Scripture should not be used to pass judgment on matters of

---

[9] See the excellent biography of Darwin by Adrian Desmond and James Moore, *Darwin: The Life of A Tormented Evolutionist* (New York: Norton, 1994), 215–216.

[10] Ernst Mayr, *What Evolution Is* (New York: Basic Books, 2001), 9.

geology, history, and archeology. For others, there were ways of reading the biblical testimony that avoided contradicting naturally collected evidence about the past.

The fossil record, in skeletal remains and imprints, is an important element in the evidence for organic evolution. One of the problems with evolution, however, is due to gaps in the evidence. This is not the same as saying that enough evidence has not accumulated yet. Nevertheless, some have mistakenly treated evolution either as unfounded or as not well thought out. Thus, some call it "just a theory." However, in scientific terminology, a theory is used to shed light upon the facts. And the facts are the evidence that we have available at any given time. Thus, a theory supported by the evidence is never "just a theory." In the almost 150 years since Darwin, facts about evolution have actually expanded to include evidence from the areas of conservation, modification, branching and structure.[11] All these processes are supported by empirical science.

Conservation is said to have happened when there was no modification in the fossil record over long periods of time. The late Harvard paleobiologist Stephen Gould and colleagues preferred to speak about "stasis," which they believed is well attested by the fossil record. Darwin thought that the gaps in the fossil evidence were due to the scarcity of well-conserved fossils, and not to actual gaps (cuts, jumps) in geological history. Nowadays, some scientists think that the gaps are a mirror to real history. Evolution may not have always proceeded so gradually and uniformly. There is evidence of brief eras (in geological terms, that is, where everything takes a lot of time anyway) of comparatively rapid population explosion and variation, as well as long periods of stability and no change. For the

---

[11] See the summary of these given by Erich Harth, *Dawn of a Millennium: Beyond Evolution and Culture* (Boston: Little, Brown & Company, 1990), 21-30.

latter to happen, it has to have been natural selection at work preserving what works well already.

Modification happens when random changes occurr and are passed on from generation to generation. Darwin himself had documented lots of cases of small variation in many populations. This leads us to branching in biology, which is the way Darwin conceived of the commonality of ancestral lines in many species.[12] This is the idea of common descent, which is so central to his theory. In recent years, advances in genetic science have given this idea of common descent quite a boost.

Finally, there is extinction, which has been documented extensively by various means, from the fossil record to genetic analysis. Extinction is the cessation of offspring or the elimination of whole populations, either gradually or catastrophically. Some biologists go as far as to say that we have information of no more than ten percent (for some it is five percent) of all species that have ever existed on earth. Extinction is part and parcel of evolutionary processes.[13]

Ever Since Darwin

The struggle for the heart of evolutionary explanations has continued to this day. Ever since Darwin, there were many questions that, because of the insufficiency of the evidence, neither he nor others could answer but only guess at an explanation. A lot more has certainly happened after Dar-

---

[12] Harth, 22–23.

[13] On this issue, see Niles Eldredge, *Darwin: Discovering the Tree of Life* (New York: Norton, 2005), 207; there Eldredge explains that paleontologists know of five global mass extinctions of organic life after the so-called Cambrian Explosion of some 535 million years ago. In all, the estimates go between 70 and 95 percent of extinction and loss of all earth's species. For Ernst Mayr, the loss through extinction is as high as 99.9 or more percent of all evolutionary lines; see Mayr, *What Evolution Is*, 140.

win.[14] It is, however, truly remarkable how many experts on evolutionary theory still appeal, or at least refer, to Darwin himself, who is still considered an arbiter on some disputes. That kind of impact is hard to find in any other scientific enterprise—especially considering that is has been a hundred fifty years since the original publication of the *Origin of Species* (1859). Besides, the so-called "Darwin industry" continues to produce a good number of studies on every imaginable facet of his life and thought.[15]

With the advent of genetic science early in the twentieth century, a new synthesis in evolutionary theory was called for. The work of researches like Ernst Mayr and the famed Russian-born biologist Theodosius Dobzhansky (1900–1975) made it possible to reach such a modern synthesis. Further studies, including the mathematically oriented population genetics, and more recently developmental biology, have challenged some aspects of orthodox evolutionary biology. In response there is a kind of neo-Darwinian theory that has dealt with and revised some core elements of the original understanding. Let us take a look at some of the most current and influential of these developments.

---

[14] A good volume to consult, from a historical perspective, is Edward J. Larson, *Evolution: The Remarkable History of a Scientific Theory* (New York: Modern Library, 2004). Another, more from the perspective of the science involved, both the disputes and possible resolutions, is the comprehensive yet tinted account by the "Darwinian heavyweight" (as Richard Dawkins has called him) Stephen J. Gould, *The Structure of Evolutionary Theory* (Cambridge, Mass.: Belknap/Harvard, 2002). For my account, I will rely on several sources, which will be acknowledged as we move on.

[15] Not least in the years leading to Darwin's 200th birthday celebration in 2009.

Descent with Modification

Evolution occurs in lineages, that is, in populations of organisms related by descent. These organisms in a population are said to be reproductively interactive in order to be considered such. In a given population, individuals share many characteristics.[16] The characteristics that are inherited, meaning, those that are passed along through generations are, therefore, similar.

However, modification (or change) happens across generations—these are changes in the frequency of the traits or characteristics of organisms. Changes in an organism during its lifetime which are not passed onto subsequent generations, are not concerned with evolution, properly speaking. As is often said, Darwin did not know the mechanism of inheritance nor of those evolutionary changes or novelties. Darwin did not know about genes, of course. He just talked about organisms being selected, and about species evolving. Since the struggle for existence happens primarily among individuals, selection is primarily working for (or against) the individual organism.[17] But the issue of how it actually happened, had to wait for further developments in the science of genetics in order to be resolved.

Nowadays, genetics dominate much of the discussion around evolution and inheritance. Genes are said to influence, although not necessarily determine, the organism's phenotype—that is, the anatomical structures, the visible physiology or even the behavior of an organism, as in the

---

[16] The discussion of this topic is indebted to the exposition by David J. Buller, *Adapting Minds: Evolutionary Psychology and the Persistent Quest for Human Nature* (Cambridge, Mass.: MIT, 2005), 17–26.

[17] See Timothy Shanahan, *The Evolution of Darwinism: Selection, Adaptation, and Progress in Evolutionary Biology* (Cambridge: Cambridge UP, 2004), 23–24.

case of what Richard Dawkins calls the "extended phenotype." It is generally agreed, however, that the phenotype does not affect the genes.

Genes can also mutate into new forms. The genetic definition of evolution asserts that evolution is change in the genes or genotypes that gets transmitted across generations. The view from the perspective of gene transformation and transmission is what is called "micro-evolution" – changes within species. But there is also "macro-evolution," which is concerned with the permanence and extinction of whole species, or with the development of new species as previous ones split or evolve owing to changes in their environment and living conditions.

For Darwin, evolutionary changes happening in individual organisms were selected by nature for their fitness or advantage to survival. What is distinctive now is that changes are seen as happening, yes, to individuals, but more exactly to the individual's genes, owing to the interaction between genes and their environment, which can lead to mutations. But we come to see the actual results of such interaction in phenotypes: physiological, behavioral, or otherwise.

All that said, the relation between genotypes and phenotypes is not really as straight or direct as the popular notion speaks of. Some genotypes can produce different phenotypes under the influence of different environmental and developmental conditions. However, phenotypic evolution is still possible in the absence of genotypic changes. It is due to environmental causes within a particular population. Needless to say, this is a debated question, whether a phenotypic difference (in the physical features of individual organisms) between two groups of plants or animals ("taxa," which is the name given to varieties or related groups within species) can be said to be caused by natural selection without changes in the genetic make up; the causes then will be almost strictly environmentally determined. Researchers who propose that it is testable have favorably argued for neutral phenotypic evolution.[18]

This brings us to the question about the causes of evolution, an issue that confronted Darwin, and for which he looked for natural explanations all the way. Again, if Darwin could not explain the causes properly, it was because as a new science his theory did not have access either to a well-developed theory of the mechanisms of genetic inheritance or to the technology that would unveil those mechanisms.

We need a cause to explain variation. In addition, we need a mechanism to keep track of the frequencies of the variations in any population.[19] One of the causes often proposed for variation is the introduction of occasional errors in the copying and replication of genetic sequences and, therefore, of the characteristics of an organism. Basically, it is to say that genes can mutate, and do mutate. Mutation is then seen as a random process, meaning, that it happens by chance or accident, and therefore, that it is not directed.[20] Which also means that not all mutations are necessarily good or to the advantage of the organism. In fact, the generality of mutations are detrimental to the individual or to the population, if spread out.

---

[18] See, for example, H. Allen Or, "Testing Natural Selection vs. Genetic Drift in Phenotypic Evolution Using Quantitative Traits Locus Data," in *Genetics* 149 (1998), 2099–2104. Or is a professor of biology at the University of Rochester in New York.

[19] Buller, *Adapting Minds*, 26–31.

[20] There is no point in denying that to insist on the strict randomness of mutations, not to say of the whole evolutionary process, is controversial. However, there is a general consensus of interpretation following this line of reasoning. That is the main reason why I refer to it for this summary at this point. In any case, it will also be acknowledged that there are those who disagree with the essence of a directionless, utterly random process, and they are not necessarily from the "intelligent design" camp.

However, the frequency and spread of a change throughout a population may or may not be as random as was originally thought. This is where "natural selection" comes in to do its job.[21] Darwin thought that for those changes provoked by mutations in the first place, which eventually prove to be beneficial to an organism and help that organism in its chances for survival, the new traits get naturally selected precisely because they were advantageous to the organism and thus to a whole population. The change is then said to help or increase the organism's fitness.

Now, the biological fitness is not given exclusively by the physical characteristic of an organism. How well an organism is adapted to its environment is key for its survival and eventual reproduction. Thus, an organism's fitness is intimately related to its environment. Yet, this relationship between fitness and environment is neither the whole nor the end of the story.

An organism that is fit enough to survive and does so, but does not reproduce (for whatever reasons) is no more fit "biologically speaking" than one that does not survive despite its potential to do so — for example, if it dies accidentally. Biological fitness is therefore related not merely to the ability or potential of an organism to survive and reproduce, but also to the actual reproduction of offspring, which carry on its genetic inheritance — or, as some call it, its "genetic contribution."[22] As happens with the appropriation by science of other terms of ordinary language, these nouns acquire a more technical definition, serving the needs of the theory, especially in its public or educational expression. Unfortunately, terms like "fitness" or "selection" cannot be completely detached from ordinary usage. Confusion of meaning or reference is almost inevitable at some point.

---

[21] Buller, 28.
[22] Buller, 29.

Biologists in general, as Darwin himself did before, make a distinction between "natural selection," which deals with problems of the adaptation of organisms and populations to their environment, and "sexual selection," which is the ability and eventual success in mating with other members of the population for reproduction. Some biologists, however, see sexual selection as subordinate to natural selection. Yet others (like paleo-biologist Niles Eldredge[23]) see sexual selection as fulfilling as important a role as natural selection, basically working together with the other mechanism for the good of the whole.

Sex is a very powerful instinct and also a drive in most animals, and all mammals. It plays a significant role, not only in reproduction *per se*, but also in the economic life — securing food and resources. Among humans, the story gets more complicated, since humans apparently use sex for other reasons too, reasons that may go beyond reproduction or economics.[24]

Debates over Adaptation and Natural Selection

Traits that appear well designed through a process of modification for solving an adaptive problem are said to be the organism's adaptation to its environment.[25] Adaptation, generally speaking, offers a solution to a problem, normally developmental in character. It may come in response to an environmental change or challenge. But even in the latter case, it can be seen as dealing with developmental problems and a needed strategy. In any case, it is essential that the

---

[23] See, for example, Eldredge's book *Why We Do It? Rethinking Sex and the Selfish Gene* (New York: Norton, 2004).

[24] Eldredge, 129–141; the author speaks of the strong triangle of relations between sex, reproduction and economics among humans.

[25] The basic definitions are given by Niles Eldredge in *Reinventing Darwin* (New York: Wiley, 1995), 33–34.

trait or property in question actually enhance the organism's chance for survival. Otherwise it would not really matter. An adaptive trait then must be well designed for life.[26]

Adaptation is the scientific accounting for the diversity of forms in the living world. From the beginning of the Darwinian theory, adaptation has been, and continues to be, the basic assumption as well as the guiding principle of evolutionary biology. It is the way in which organisms match their environments in order to thrive. Basically, it is a passive filter that allows features of an organism to pass along to the next generation.

Natural selection tends then to retain those modifications that are beneficial to individuals and populations through time. It is this cumulative retention of (for the most part) favorable modifications that hints toward design. Darwin himself at times made natural selection sound like a designer or artificer. But, ultimately, natural selection is blind to specific outcomes. Only this can be said: what are passed on are those features that worked better than others in the previous generation.

One complaint has been that, in the evolution and adaptation of complex structures (the eye, for example), evolutionary biologists have given too little attention to intermediate stages. This argument has been exploited by so called "scientific creationists" to point out supposed flaws in evolutionary theory. Their main point of argumentation seems to be that complex, irreducible structures are a whole system, indivisible as to their concrete function, which if divided by individual staged adaptation, stop making sense. By themselves, none of the stages of its supposed evolutionary history would work. It is argued then

---

[26] See the explanatory remarks on this point by Mark Ridley, "The Mechanisms of Evolution," *An Evolving Dialogue: Theological and Scientific Perspectives*, ed. James B. Miller (Harrisburg: Trinity, 2001), 53.

that complex structures are irreducible to their parts and, therefore, need to have been specifically designed for their current function or role.

Evolutionary biologists in general have insisted that to think of evolutionary stages as pieces of a big puzzle, which need to be put together or would always be incomplete, is not how they think of the different adaptations required along the history of that appropriate organism or trait in question. At different moments of the organism's development, different physiological traits were used differently: they had different functions either related or unrelated to their current, but not necessarily "final," product.

In the 1970s, Stephen Gould and a colleague of his[27] launched a strong critique against what they termed the "adaptationist programme" in evolutionary biology. Their basic argument is that not everything in the evolution of traits or characteristics is a solution to an adaptive problem. Quite a few physiological developments happened for different reasons: environmental, accidental, chance mutation, and did not necessarily fill a niche in evolution. However, a few of those have eventually been co-opted by natural selection because of a specific advantage conferred to the individual organism or population at a given moment of their history.

Gould, together with another colleague of his, Elisabeth Vrba, coined the term "exaptation" to be able to speak about such features. Some were, so to speak, "left over" from previous developments. That is where the image "spandrel" becomes handy for Gould et al. As in architecture and art, spandrels are structures not thought of in advance, not spe-

---

[27] Stephen J. Gould and Richard C. Lewontin, "The Spandrels of San Marco and the Panglossian Paradigm: A Critique of the Adaptationist Programme," *Proceedings of the Royal Society of London*, Series B: *Biological Sciences* 205 (1979), 581–598. This article has been reproduced in different publications, given its seminal importance for the debate.

cifically designed or planned, but rather an unintended consequence of some other designed structure.[28]

Likewise in evolutionary history, there are instances of structures that either were "spandrels" from other developments, or that were originally adaptations to specific living conditions and, when those conditions were no longer, those very structures were used or adapted in other useful ways. Gould thinks of language as one such "exaptation." In the latter case, after the human brain had grown at least double its size in comparison with other hominids, language was developed—which is not to say that language was the reason why the brain grew to its current size in the first place.

In general, evolutionary biologists have accepted the above point. Most of them attest to the complexity of the evolutionary process. There is a lot to be said. Where they differ is in the assessment that Gould's criticism may necessarily imply the abandonment of the search for adaptive explanations or of giving proper rationale to surviving adaptations. As a matter of fact, evolutionary psychologists seem to have imposed on themselves the task of finding explanations for all kinds of possible adaptations, especially as these may have influenced human behavior, instincts or motives, from ancient times to present circumstances. And they do so in the belief that it can actually explain a lot about current cultural problems, behavioral malaise or communal disruptions in modern life.

Among the cases often cited as presenting problems to an adaptationist explanation in biology, are the function of male nipples and the development of the female orgasm. It is normally agreed that male nipples have no clear use,

---

[28] "Spandrel" is the space left by arched columns that support a semi-cannon dome, as in quite a few European cathedrals, "San Marco" in Venice, being one such good example, especially since the spandrels in there were covered and painted over through time.

despite some attempts at finding some adaptive explanation, i.e. for touching and arousal that may increase the chances of reproductive success — a bit of a stretch. But the generally accepted explanation is that male nipples are the result of a parallel embryological development, that is to say, that males have them because women have them too and needed them in the first place. A parallel anatomical structure develops because the embryological stages among different organisms are very similar if not identical.

When it comes to female orgasm, it is also conceivable to think of it as a byproduct of embryological development. However, in contrast with the example from male nipples, in the case of the female orgasm, a good number of diverse evolutionary accounts — twenty-one by one count — have been proposed.[29] According to biologist and historian of science Elisabeth Lloyd, research on female orgasm has proven most of these explanations insufficient, if not faulty. In addition, female orgasm is a good case study of bias in evolutionary science. It has been assumed, for instance, that female orgasm contributes to the reproductive success of women. Those looking for an adaptationist explanation — what she calls the bias of adaptationism — have favored this argument. Moreover, it has been assumed that female sexuality is like male sexuality. But the evidence points in a different direction: that men and women have dissimilar sexual responses.

For Lloyd, the adaptionist explanation simply fails in this case. Of the twenty-one explanations given for female orgasm, only a non-adaptive one actually explains something. Female orgasm does not seem to qualify as an adaptation, although it is still the result of an evolutionary development, because it does not seem to pass a rigorous

---

[29] See Elisabeth A. Lloyd, *The Case of the Female Orgasm: Bias in the Science of Evolution* (Cambridge, Mass.: Harvard UP, 2005), for a well-researched and, in my opinion, convincing argumentation on this regard.

test of the evidence for adaptation.[30] Despite the many explanations offered, the evidence remains inconclusive. In any case, and this is one of Lloyd's contentions, the sort of explanation favored by one researcher or another seems to be influenced by factors other than the evidence alone.

Is Natural Selection It?

There is no doubt that natural selection is the core principle of Darwinian evolution. But, is natural selection the only determinant in the evolutionary process? For Darwin, it was certainly the main mechanism but not the exclusive one. Chance or accident, changes in the environment, and the occasional ecological catastrophe play important roles in the process of evolution, too.

Natural selection is neither an engineer, one that proceeds by designing new structures or modifying old ones or creating new models, nor an experimental lab scientist, testing and probing good designs and discarding bad ones. Actually, some adaptations can be detrimental either in the short or in the long run, especially the latter case.[31] New forms that appear in the evolutionary process are not necessarily aimed at improving functions, or at solving specific problems. It is evolution without a pre-conceived plan or purpose. And this seems to be what adepts of some religious or theological views find the most difficult to accept. First, because it sounds like religious dogma itself. Second, because questions of purpose or non-purpose are probably equally untestable, if we follow the same scientific rationale for both assertions. In any case, it seems to be among the reasons and motives behind the search for "intelligent design" arguments in science and religion alike.

---

[30] Lloyd, 4–7.
[31] Harth, *Dawn of a Millennium*, 27.

The point that needs to be underscored is that Darwin was against any forms of essentialism. For example, a population is not a group of "types" (to which every individual conforms) but a population, precisely, of individuals. The "average person" is an abstraction. Of course, we all share many characteristics or traits, but each individual is also unique in a sense. Scientists now know for sure that we each have our unique genetic pattern or blueprint when it comes to individual details.[32] Life is not just diverse in the general, say, among species; it is equally diverse within each population, family and group, down to every individual.

The Case for Evolutionary Pluralism

As it stands today, evolutionary theory is basically an assembly of scattered facts, together with some deductions, a few conjectures, and many debatable or controversial opinions.[33] Evolutionary pluralism is more the case than not. Heterodoxy seems to loom large and growing somewhat quietly but firmly. Evolutionary orthodoxy, which is normally the term applied to those who hold a restricted view of Darwinian mechanisms as sufficient to explain basically all questions in biology, gets more combative, aggressive, and at times, rigid. Although not by any stretch evolutionarily orthodox , the late Theodosius Dobzhansky (1900-1975)[34] is often quoted by many an evolutionist with

---

[32] Harth, 23.

[33] Harth, 29.

[34] Dobzhansky was a pioneer in the study of population genetics, which serve as a tool to better explain how the variety of groups, races, and species happens through natural selection. But Dobzhansky was also a committed Orthodox Christian, who did not hesitate to openly express his religious and humanistic views as compatible with science. His main work on this area is *The Biology of Ultimate Concern* (New York: The New American Library, 1967).

his well-known statement that "nothing in biology makes sense except in the light of evolution."[35]

Generally speaking, evolutionary theory lacks the neatness and compactness of other theories in science, for example, Newtonian gravitation, Huber's cosmology, or Einstein's relativity. It is not easily wrapped in mathematical equations—though it has been tried. However, it does not mean that evolutionary theory is necessarily flawed. It is just incomplete: it is an open system, an exercise in inductive thinking, but still with plenty of evidence to study and from which to draw further conclusions.

Where does natural selection happen? What is the unit of selection, properly speaking? By the 1960s, after a century of studies since the original publication of Darwin's *Origin of Species*, evolutionary biologists were ready to re-visit this issue. First it was George Williams (b. 1926)[36] who came out against the idea that selection may happen at the level of groups—an idea previously advanced by those who worked the "Darwinian synthesis" a few decades before—and against the notion that selection is not necessarily for the good of the individual, a notion that went back to Darwin himself. These were in fact two challenging ideas that provoked a set of new studies and controversies. Williams is mostly known for his contribution to a gene-centered view of selection in evolution, so influential in subsequent evolutionary studies. In any case, what matters now is to say that people like Williams were trying to resist any conception of direction or design in evolutionary thinking.

---

[35] From the title of an article by Dobzhansky, "Nothing makes Sense in Biology Except in the Light of Evolution," in *The American Biology Teacher* 35 (1973), 125–129.

[36] See George C. Williams, *Adaptation and Natural Selection: A Critique of Some Current Evolutionary Thought* (Princeton: Princeton UP, 1966). Richard Dawkins has acknowledged Williams as one of his major influences.

According to Eldredge,[37] however, by insisting that natural selection works only for the good of the individual, they had made it into an active force, more powerful than it was ever conceived within traditional Darwinian circles. In fact, they differ from Darwin's own approach, while claiming ironically to hold firmly to Darwin's original views.[38] What is at stake here is nothing less than the nature and understanding of "purpose" in biology. It rings of religious motifs, which makes many scientists uncomfortable. As mentioned before, the reference to "purpose" becomes like a charge of "heresy" to throw at opponents.

For their part, population geneticists re-conceptualized the term "fitness," which originally meant simply "reproductive success" in biology. They expanded it to include the success of individuals within a population to leave as many copies as possible of their genes to the next generation—or "relative fitness," which is different from Darwin's notion of fitness as individual vigor to reproduce.

Overemphasis on reproduction over, say, economic success (viz., securing food) to live, becomes then the sole purpose or goal of the individual organism. The latter is conceived almost exclusively as being in active competition against other individuals, especially since the success of one could actually mean failure or danger to another.

It was not difficult to see the next step in this narrowing of the concept of selection. It happened with the publication of

---

[37] Eldredge, *Reinventing Darwin*, 33.

[38] Appeals to Darwin himself are rather common in evolutionary controversies. It speaks in a powerful way to the iconic presence and the deep impact that Darwin caused upon the science of evolution, a development practically unparallel in other scientific disciplines—although in non-scientific ones (despite other claims), say, in psychology, Freud seems to represent a parallel point of reference. In any case, this may be considered a trait of religious fervor, that is, a way of making past personages into "holy icons" or at least as arbitration authorities.

Richard Dawkins' *The Selfish Gene* (1976). For Dawkins, there are two basic kinds of entities: replicators and vehicles (a distinction that was further clarified in his subsequent volume *The Extended Phenotype* (1982).[39] According to Dawkins, genes are replicators, but need vehicles for their survival. Furthermore, those vehicles are basically conduits that allow genes to replicate and further their reproductive activities. The genes are, therefore, the ones competing for survival. Life in this picture is all about reproduction. The genes are then basic and minimal units of selection. The rest is commentary, so to speak.

However, for others (like Eldredge) life is primarily about securing the means for survival. Reproduction comes after the fact, "not so much as an imperative but as a physiological luxury."[40] Moreover, there are those who pose changing views about evolution and heredity, challenging the "gene-centered" version of neo-Darwinism, which has been dominant for almost five decades (since the inception of the "new synthesis"). One such contribution comes from the likes of Eva Jablonka and Marion Lamb, both of whom expand their understanding of evolution.[41]

First, by insisting that not all heredity is gene-originated, then Jablonka and Lamb explain that there is more to heredity than genes. There is, second, epigenetic inheritance, through which some acquired information is also inherited. Third, the assertion that evolutionary change can be better understood as a result of both instruction and selection—and not natural selection alone. Fourth, there is evo-

---

[39] Richard Dawkins, *The Extended Phenotype: The Long Reach of the Gene* (revised ed.; Oxford: Oxford UP, 1999).

[40] Eldredge, *Reinventing Darwin*, 40.

[41] Eva Jablonka and Marion J. Lamb, *Evolution in Four Dimensions: Genetic, Epigenetic, Behavioral and Symbolic Variation* (Cambridge, Mass.: MIT, 2005). This work has been well received especially by those who hold a pluralistic view of evolution.

lution that occurs through the transmission of behavior, in which the development of culture (which is not exclusively human) plays a significant role.

According to biologist Massimo Pigliucci, the work of Jablonka and Lamb is a contribution to a new neo-Darwinian synthesis, which is also critical of the shortcomings of the previous generation of evolutionary theorists.

> There has been rumbling for some time to the effect that the neo-darwinian synthesis of the early twentieth century is incomplete and due for a major revision. In the past decade, several authors have written books to articulate this feeling and to begin the move towards a second synthesis. David Rollo, in his book *Phenotypes* (Kluwer, 1994), was among the first to attempt to bring the focus back to the problems posed by phenotypic evolution. In *Phenotypic Evolution* (Sinauer, 1998), Carl Schlichting and I framed the debate in terms of the integration of development, environment and genetics by articulating the concept of "developmental reaction norms". Stephen Jay Gould then produced an overly long (and at times acrimonious) sketch of the new synthesis in *The Structure of Evolutionary Theory* (Harvard University Press, 2002). Finally, Mary-Jane West-Eberhard, in *Developmental Plasticity and Evolution* (Oxford University Press, 2003), greatly expanded on my book and the one by Rollo, producing the most comprehensive alternative account of evolutionary theory yet. *Evolution in Four Dimensions* by Eva Jablonka and Marion Lamb is the most recent addition to this genre, and contributes yet another valuable perspective to the discussion.[42]

We can safely say that Jablonka and Lamb's contribution falls within the camp of "Darwinian pluralism" — which stands *vis à vis* the stance of a kind of "Darwinian fundamentalism" — a pluralism about which the late Stephen Gould (1941–2002) wrote so much and for which he became

---

[42] Massimo Pigliucci, "Expanding Evolution," *Nature* 435 (2005), 565.

an apologist.[43] For Darwinian fundamentalists, natural selection is all there is to explain the diversity and complexity of life. Pluralists, on the other hand, seek to understand a number of interacting principles, natural selection being primary among them.[44] In the latter case, evolutionary explanation is not reduced to one single and overarching principle. As we can see, the major issue is not evolution itself but its mechanisms.

In Gould's (and others') interpretation, Darwin himself had a pluralist view about such mechanisms. Darwin's often-quoted phrase in this regard states that "natural selection has been the main but not the exclusive means of modification,"[45] opening the door for research of parallel, even if in comparison lesser, mechanisms. Contingency, for example, is also a player.

> Crank your algorithm of natural selection to your heart's content, and you cannot grind out the contingent patterns built during the earth's geological history. You will get predictable pieces here and there (convergent evolution of wings in flying creatures), but you will also encounter too much randomness from a plethora of sources, too many additional principles from within biological theory, and too many unpredictable impacts from environmental histories beyond biology (including those occasional meteors) — all showing that the theory of natural selection must

---

[43] For example, Stephen J. Gould, "Darwinian Fundamentalism," *The New York Review of Books* 44/10 (June 12, 1997), 34–37.

[44] S. J. Gould continued his arguments in a follow-up article, "Evolution: The Pleasures of Pluralism," *The New York Review of Books* 44/11 (June 26, 1997), 47–52; unfortunately, in both installments Gould utilized too much writing to defend himself against (what he considered) Daniel Denett's personal attacks, as well as to his critique of evolutionary psychology.

[45] The quote is given by Gould in his essay on pluralism, page 50 (see previous note), and taken from the autobiography of Charles Darwin.

work in concert with several other principles of change to explain the observed pattern of evolution.[46]

Moreover, Gould has insisted that selection itself has acted at multiple levels, not merely on individual genes, which has been dogma especially since the "new evolutionary synthesis" was put together a few decades ago.[47] But selection also works on populations and species and even on ecosystems as a whole.

Pluralism is not just a matter of interpretation in evolutionary biology. According to Mayr, pluralism is characteristic of the evolutionary process as a whole. He calls evolution an "opportunistic process," one that takes advantages of present circumstances, or whatever means are available, in order to better accommodate and therefore accomplish the needed results.[48] For Gould, however, the bottom line is complexity: "We live in a world of enormous complexity in organic design and diversity — a world where some features of organisms evolved by an algorithmic form of natural selection, some by an equally algorithmic theory of unselected neutrality, some by the vagaries of history's

---

[46] Gould, "Evolution: The Pleasures of Pluralism," 48.

[47] The term "new or modern evolutionary synthesis" is used for the concilience of Mendelian genetics and Darwinian natural selection in the early decades of the twentieth century. Eventually it also included mathematical population genetics. The names most often related to the early "synthesis" are those of Thomas Morgan, Ronald Fisher, Theodosius Dobzhansky, Ernst Mayr, and George Simpson. In the latter period, we have William Hamilton, Gaylord Simpson, George Williams, John Maynard Smith and, more recently, Richard Dawkins. It is not a monolithic movement, if a movement at all. And it cannot be described as just a stage in evolutionary thinking, since it has taken decades to developed and it is pretty much where evolutionary biology finds itself today, despite of revisionist efforts and new pursuits in biology.

[48] Mayr, *What Evolution Is*, 221.

contingency, and some as byproducts of other processes."[49] Gould left the door open to other possible, albeit minor, mechanisms at work in evolution.

The Challenge to Darwinian Gradualism

Evolution of living forms is a gradual process, by the slow accumulation through time of small changes. However, the fossil record has many gaps: it is discontinuous. The major gaps seem to happen in geological strata marking transitional periods. However, the morphological changes in living forms, according to the available evidence, are not necessarily radical: forms seem to persist for periods representing millions of years.

There are at least two possible interpretations for this seeming persistence of forms through time. One possibility is that the accumulation of sediment in geological strata is itself uneven, probably owing to climate changes or other geological changes or disasters, e.g., earthquakes and meteorites. The other possibility is that there has always been a form of evolution that does not progress uniformly or evenly. There may have been long periods of time with little or no evolutionary change and also periods of rapid change (by the standards of geological time, that is) in forms and the eventual evolution of whole species of organisms. Sometimes the formation of new species seems to happen in sudden bursts of creativity.

This process has been termed by Gould and Eldredge the "punctuated equilibrium" model of evolution. As a theory, it proposes that a good number of plant and animal species have arisen relatively quickly in terms of geological time (in less than 100,000 years) and not so much through a process of gradual change, at least not as has been traditionally understood. Since its conception, this idea of "punctuated

---

[49] Gould, "Evolution: The Pleasures of Pluralism," 52.

equilibrium" has been presented as a corrective to an important tenet of Darwinian theory.[50]

However, to this day, whether evolution has been predominantly punctuated or gradual has remained an open topic of debate. The question then is not so much whether evolution has occurred or not, but about the frequency and pattern, if any, of evolutionary changes. It may after all be in the nature of a gradual evolutionary process to have the rate of change vary from time to time.[51] Nevertheless, gradualism has normally been equated with evolution itself. The point of the "punctuated equilibrium" theory has been to validate the paleontologists' data instead of claiming—a trend starting with Darwin himself—that all is due to voids in the evidence. According to Gould, the theory has been misunderstood for a "saltational interpretation,"[52] which had been proposed from the earliest periods of evolutionary theory as the radical and quick transformation of a number of species.[53] Darwin himself had already denied any validity to such a view. He saw evolution as a gradual and successive process, so infinitesimally small as for us not to be able always and continually to detect it in the fossil record.

---

[50] Stephen Gould and Niles Eldredge introduced the concept of "punctuated equilibrium" originally in a co-authored paper "Punctuated Equilibria: An Alternative to Phyletic Gradualism" in 1972. It has been since reproduced many times, for example, in T. J. A. Schopf, ed., *Models in Paleobiology* (San Francisco: Freeman, Cooper and Co., 1972), 82–115.

[51] For a good though brief summary of these points, see Francisco Ayala, "Biological Evolution: An Introduction", *An Evolving Dialogue*, especially 44–45.

[52] "Saltationism" is the belief that the evolution of variations in organic life has proceeded in "jumps through the inheritance of gross mutations." See Larson, *Evolution*, 87.

[53] Stephen Jay Gould and Niles Eldredge, "Punctuated Equilibrium Comes of Age," *An Evolving Dialogue*, 167–180.

The paleontological evidence shows that stability rather than change dominates the fossil record. The rate of "stasis" in a number of species is actually measurable.[54] What this proves is the general stability of species through time. For Gould, this stability can be seen as an active phenomenon, a regular tendency of species to remain unchanged, at least as long as a change in the environment or in the living conditions does not demand otherwise.

This challenge to gradualism is not a foregone conclusion to many evolutionists. The theory that asserts periods of stability followed by periods of sudden change has its own difficulties, which remain unanswerable for the most part. For example, Ernst Mayr has remarked that what Darwin understood by gradualism was not the same as uniformity in the rate of evolutionary change. Moreover, the fossil record that is used to make the case one way or another is not exclusively depended on local changes in a population, but rather on the geographical spread of such a population. Finally, Mayr and others have pointed out that variation happens more frequently in smaller populations, and these are harder to be registered in the fossil evidence.[55]

Another difficulty confronted by punctuated equilibrium has been academic and at times political rather than scientific, meaning, it calls for a re-evaluation of the "new Darwinian synthesis," an attempt not really popular among a few evolutionists. However, Mayr has said that the theory does fit the facts rather well, and it still is the best inference that evolutionists have now.[56] It is my opinion that the cred-

---

[54] Gould and Eldredge, 169.
[55] Both points are made by Mayr in "Speciational Evolution and Punctuated Equilibria," in *The Dynamics of Evolution* (ed. Albert Somit and Steven Peterson; New York: Cornell UP, 1992), 21-48. Mayr is celebrated as one of the greatest evolutionist of the 20th century, and therefore very much respected for his opinions.
[56] Mayr, 33-34.

ible aspect of the theory is to think of genetic variation as a cumulative progress to the "tipping point."[57] But, it is one thing to think in terms of the preservation through time of hard structures, and another to speak about the rate of molecular or behavioral changes. For the last two, the paleontological record is simply of no help.

Evolution and Emergence

The notion of a tipping point of progression that can be used in the sciences is the popularization of another concept in science, that of "emergence." For the last two decades, there have been a few studies on how a conception of emergent systems may shed light on the evolution of novel forms, especially in the appearance of complex systems, biological, chemical or otherwise physical. I for one find the discussion of emergence theory to be very helpful in dealing with questions in evolution, in spite of the incompleteness of our understanding of the origins and evolution of life, or rather precisely because of the complexity of the questions that still remain unanswered.

What is emergence? According to philosopher of science Philip Clayton, defining emergence is not easy because it is not a monolithic term. In its technical sense, as it is applied in science, "emergence is the theory that cosmic evolution repeatedly includes unpredictable, irreducible, and novel

---

[57] The accumulation of little changes through time can result in big changes. After a period of sustained repetition, and at the point of reaching "critical mass," can cause an avalanche of changes, transformation, novelty or, plainly, growth. The "tipping point" effect is being applied—sometimes a little indiscriminately—to multiple phenomena, in science, math or pop-psychology. See, for example, Malcolm Gladwell, *The Tipping Point: How Little Things Can Make a Big Difference* (New York: Back Bay, 2002).

appearances."[58] Emergence is often explained as more than the sum of the parts of any system, not identifiable with any part exclusively and yet not other than the parts. In any case, emergence is used to oppose reductivist positions in science.

Emergence theory is not really a newcomer. In the early decades of the twentieth century, a few scientists and philosophers of science, who were deeply influenced by evolutionary thinking, began to deal with the question of emergence of new systems and life forms in nature. Emergence was initially considered as an alternative to Darwinian gradualism in biology.[59] Emergence is the attempt to speak of the importance of qualitative novelty in evolution.

An early example of this concern is the work of Conway Lloyd Morgan. Morgan's aim was to consider the possibilities for the appearance of new life forms through natural processes and without recourse to direct divine intervention. In addition, he advocated the analysis of reality at different levels, with the intention of looking for novel forms or behaviors in nature. This was his way of looking for novelties that were not parts originally of a given natural system. He sought for ever-increasing levels of complexity in the world of organic forms. The problem with Morgan's ideas is that they require strong ontological commitment, that is to say, emergent qualities as Morgan thought of them, are more than mere patterns.[60] They actually seem like substances or essences, ontologically speaking. But such assumptions or consequences of a theory are not for the most part well received in science. Science is said to thrive on parsimony of explanation, meaning, that no more beings or qualities or forms than is strictly necessary ought

---

[58] See, Philip Clayton, *Mind and Emergence: From Quantum to Consciousness* (Oxford: Oxford UP, 2004), 39.

[59] Clayton, 13.

[60] Clayton, 13-14.

to be assumed or introduced in order to explain something.[61]

Philip Clayton sees emergence as an ontological alternative to the extreme options of physicalism and dualism in science and philosophy.[62] For him, those are the two alternatives that the tradition of natural philosophy has bequeathed to us. Emergence theory, then, is a way of trying to break out of these reductive interpretations of reality. The study of emergence is a way of assessing creativity and novelty in an old universe. It is a kind of "bottom up" exercise, where simple features of reality (e.g. atoms) are followed in their complex interactions, as they lead to the formation of new units of reality—from atoms to mind, as Harold Morowitz likes to say.[63] What appears before us is the emergent nature of an evolving universe.

In the realm of research disciplines—say, for instance, physics and mathematics—hypotheses and formulas are constructed sequentially, one building upon the other, either assuming the validity of the ones we have at hand, or by trying to correct and expand them so as to allow them to produce novel results. The same applies to the way that scientific disciplines interact: cross-fertilization among them allows for new perspectives or explanations to develop. This is also an emergence view.

All that said, there is a kind of epistemological circularity at work here. One starts with an assumption of a "primitive reality" (as Morowitz calls it) in order to explain a more complex one, which is assumed as an end result of the former.[64] But emergence is more than a way of describing the

---

[61] Clayton, 17.
[62] Clayton, 38.
[63] Harold J. Morowitz, *The Emergence of Everything: How the World Became Complex* (Oxford: Oxford UP, 2002), 1–14.
[64] Morowitz, 8.

development strictly of physical systems. It moves beyond materialist explanations.

In biology, for instance, it is common nowadays to say that science deals, not only with matter and energy, but with information as well. Some researchers, like the British biologist Steven Rose, call for an "integrative biology," which may be in a better position to explain the complexity of organic systems while using, at the same time, what he calls "epistemological diversity" in the search for the nature and the meaning of life. Rose believes that bottom-line life enjoys some kind of ontological unity.[65]

It is probably in computer technology that emergence theory has had the greatest impact, and this impact has again spilled onto the natural sciences. Computers allow scientists to tackle difficult problems and take them to a greater level of complexity by dramatically increasing their calculation capacities. With the significant increase in the accumulation of data and calculation powers, traditional fields of investigation as well as new systems of ideas demonstrate emergent capacities for complexity and analysis and therefore for explanation. The computerized models for complexity that have been developed, for example in chaos theory, have already been applied to the different sciences. In biology, for example, through the use of mathematical and computer models to explain evolution, with different levels of accomplishment and degrees of success.[66]

---

[65] See Steven Rose, *Lifelines: Life Beyond the Gene* (revised ed.; Oxford: Oxford UP, 2003); Rose, who is a prolific writer, is one of the most outspoken scientists against forms of determinism and reductionism in science, which he argues make ideological besides scientific claims, especially in biological and genetic research.

[66] A good example of this kind of work is carried by the likes of Stuart Kaufman at the famed Santa Fe Institute, which specializes in multidisciplinary approaches to science, social science, and technology, and well-known for their studies of

In some cases, the complexity is so great that the problem surpasses present calculation capacities.[67]

A reductionistic approach, on the other hand, looks into lower hierarchical levels. "Emergence then is the opposite of reduction. The latter tries to move from the whole to the parts. It has been enormously successful. The former tries to generate the properties of the whole from an understanding of the parts."[68] The point is that there are emergent properties of systems that have to be understood as properties of the system in question as a whole. A mere analysis of the individual parts, although important in itself, does not necessarily convey a good understanding of the respective system. This, by the way, is also precisely one of Midgely's strongest arguments. For her, the idea that only parts, units, or individuals are real is misleading. It is bad metaphysics. "Wholes and parts are equally real."[69] Moreover, when the parts come together, they work together, and new levels of experience and reality are manifested. In brief, wholes matter, whether we are talking about the earth as a living system, of our selves as living persons, or communities as organic entities.[70]

Emergence theory is helpful in evolutionary theory, but not because it necessarily proves or disclaims the tenets of evolution. Emergence helps to identify emergent elements or properties in organic systems, for instance, in the biosphere, as well as other emergent features of the natural

---

chaos and complexity theory. See Stuart Kaufman, *At Home in the Universe: The Search for the Laws of Self-Organization and Complexity* (Oxford: Oxford UP, 1995).

[67] Morowitz, *The Emergence of Everything*, 13.
[68] Morowitz, 14.
[69] Mary Midgley, *Science and Poetry* (London: Routledge, 2002), 186.
[70] Midgley, 16.

world.[71] In evolution, novelty occurs by mutation, selection and survival—new organic structures or systems emerge that cannot be said to have been designed or predicted in any specific way. In any case, emergence leads to novelty, which (again) implies that the new forms, system or reality, are not merely the sum of its parts.[72]

The Evolution of Evolution

Evolutionary theory continues to evolve, with new developments in scientific methodology, and new contributions from paleontology, biochemistry, molecular biology, genetics and other related disciplines, many of which were not available to Darwin. But the general understanding both within and without those fields is that the basic tenets of the theory have been confirmed beyond any doubt. From the elaboration of the so-called "modern synthesis", with the integration of genetics, in the 1930s, to its deepening in the 1950s with the discovery of the structure of DNA, and to its current broadening with the addition of pluralistic interpretations, further evidence has accumulated and strengthened the theory's detail and persuasiveness.

Since Darwin, the evidence for evolution has been provided on four basic fronts. For example, the fossil record has provided strong evidence for modification of life forms through time as being consistent with common descent. The evidence in this regard has been very strong to this day, with no proof to the contrary. There have been no reversals, say, such as the appearance of complex forms before simpler ones, that have been found in geological strata.[73] Moreover, the study of comparative anatomy has offered significant clues about common ancestry among many

---

[71] Clayton, *Mind and Emergence*, 85.
[72] Morowitz, *The Emergence of Everything*, 20.
[73] Ayala, "Biological Evolution," 21.

animals. Biogeography, the science that studies the distribution of organisms, past and present, through spatial patterns of biodiversity, is also able to account for the extraordinary diversity of life. It does it by tracing patterns of variation of living things over the earth.[74]

In the case of molecular biology, the evidence for evolution has been strengthened by showing common ancestry between all living forms, from bacteria to humans. Organisms show a great uniformity at the molecular level, in the way that organisms are built—that is, their molecular structures. For example, in all organisms, DNA is assembled with the same four basic nucleotides, which attests to the genetic continuity of all life. This is additional evidence of the common origin of all living forms.[75]

Evolutionary biologists had basically abandoned the study of embryology early in the twentieth century, although Darwin himself had cherished it much and referred to in his *Origin of Species*. However, embryology has been revived lately as the source of new studies on evolutionary development (also known as the science of "evo-devo"). Nowadays there is a renewed attempt to understand how the development of forms, from embryonic to adult stages, affects the course of the evolution of organisms. As mentioned before, the study of parallel anatomical structures is but one example of such efforts.

In other developments, and out of the new batch of studies, what are actually remarkable (at least according to Michael Ruse[76]) are the similarities in DNA material between many species, from insects to mammals, including humans. The question is still open as to whether these

---

[74] See Mark Lomolino et al., *Biogeography* (3rd ed.; Sunderland, Mass.: Sinauer, 2005), 3.

[75] Ayala, 28.

[76] Michael Ruse, *The Evolution-Creation Struggle* (Cambridge, Mass.: Harvard UP, 2005), 193.

"DNA homologies"[77] are stronger evidence, and stronger players, than natural selection itself in the evolutionary process. Natural selection is being looked upon more and more as basically a filter for unsuccessful morphologies generated by organic development. But for others, phylogeny is not an alternative to selection but simply additional evidence of the richness of the evolutionary process — not an argument against the former.[78]

True to the core idea of evolution as a continuing process, subject to study and therefore knowable (at least to a certain extent), yet unpredictable due especially to the contingent nature of any historical process, many are cautious about how they conceive of it. Pluralistic understandings are increasing, while questions about the philosophical assumptions that seem to have accompanied many proposed solutions in the past, call for a revision of the received scientific wisdom up to this point. There is no reason for treating evolutionary theory itself as unchangeable or finalized or absolute in any conceivable manner. As theologian Gerd Theissen writes,

> Those who accept [the theory of evolution] as a paradigm of contemporary thought have no reason to make it absolute: evolutionary theory, too, has undergone evolution. It corresponds to the present state of our knowledge and our mistaken ideas — and no more. It is a structure by which our knowledge can adapt to reality (and probably has only limited validity). But it is one of the most fascinating construc-

---

[77] Features of organisms that have the same evolutionary origin but have developed different functions are described as being "homologous" to one another. Moreover, structures such as chromosomes and DNA molecules are likewise said to be homologous when they have basically the same structural features. See for example *A Concise Dictionary of Biology* (2nd ed.; Oxford: Oxford UP, 1990), 119.

[78] Ruse, *The Evolution-Creation Struggle,* 193–194.

tions of human reason, an attempt to give an explanation of the framework which determines our life.[79]

Darwinian theory cannot be expected to explain every conceivable question concerning life's origins and development. As a scientific theory, it is the product, not just of the evidence but also of the state of human knowledge at this point in time and place. It is a historical development and very much influenced by social context, like any other form of provisional explanation. Nevertheless it is also true that evolutionary theory is still the most reasonable inference and the best possible explanation that fits the available evidence.

Needless to say, not all questions in evolution have been solved. Of special interest is the problem of the origin of life. Despite the many proposals or possible solutions to this "mystery of mysteries," we are not any closer to anything like an agreement. Part of the issue is that, as scientific ideas and research stand now, the solution to the problem requires greater knowledge of biochemistry than evolutionists normally have. But even in this area, further studies in RNA (and not just in DNA) look promising. It may help to untangle complex questions around cell development.

Evolution and Public Perceptions

Evolutionary theory is not merely intra-scientific talk; it is very much present in the public arena. This is not merely due to the ever-present debate between teaching evolutionary and creationist views in public schools. It is also due to the presence and impact of bona fide scientists in the public media, through best-selling books, public station documen-

---

[79] Gerd Theissen, *Biblical Faith: An Evolutionary Approach* (trans. John Bowden; Philadelphia: Fortress, 1985), xi.

taries, and even science-oriented movies in the theaters.[80] At least in the United States of America, a week does not go by without some reference in the news about developments on the so-called "evolution-creation debate," or a few articles on some new developments, research or ideas in evolutionary science. In general, science is a regular topic of discussion on both public and commercial media.

Among the fears and uncertainties that the teaching of evolutionary biology arises in the public, none is more controversial than the perception that the theory is essentially antithetical to Christian faith and values in particular, and against religion in general. Although it does not necessarily have to be this way, there is yet some justification for the resilience of these sentiments. Either because of the propaganda efforts from a few openly atheist scientists (Richard Dawkins and Peter Atkins being two good examples), or because of the tough questions that a philosophy of evolution raises about the origins and roles of values (including faith and morals), the science of evolution has been thought as being unavoidably materialistic or even outright anti-religious. Moreover, evolution is sometimes presented as a kind of religion. Philosopher of science and religion Michael Ruse agrees that,

> In England and even more so in America... Darwinian evolutionary biology continues to function as a kind of secular religion. It offers a story of origins. It provides a privileged place at the top for humans. It exhorts humans to action, on the basis of evolutionary principles. It opposes other solutions to questions of social behavior and morality. And it

---

[80] Good examples from the wide screen are the movies "March of Penguins", and "What the 'Bleep' Do We Know?" shown in theaters across the country. The former deals with evolution and the biology of cooperation, the latter with questions of quantum mechanics and perceptions of reality.

points to a brighter future if all is done as it should be done, in accordance with evolutionary theory.[81]

When then is evolution a religion, or religious? Are evolutionary and religious views on life necessarily antithetical? To change the question a bit, what can religious thinking, or theology for that matter, learn from evolution as a science of nature and of the human? These are some of the questions that we would like to explore in the next chapter.

---

[81] Ruse, *The Evolution-Creation Struggle*, 213.

Chapter Four

Travails of Evolutionary Theory with Religion

As I hope will be clear through these pages, evolutionary theory is not necessarily incompatible with religious belief. On this matter I side with Mary Midgley when she states that one of Darwin's major contributions to philosophy, and I would add to theology as well, is the conviction that we belong down here, that we belong to the earth, that we are part of creation, not under or above the whole of the biosphere.[1] Darwin's common sense has brought our attention back down here from up there. Most importantly, when the theory is properly assessed, it becomes quite a corrective to human arrogance, and to any religion that forgets where our proper place is: down here, with every other creature. In this sense, the theory could be said to be able to contribute to a spiritual if not a religious view of how things really are, us included.

For Midgley, science, especially biology, can be considered a source of both knowledge and understanding. We need science, and also more than science, on our way to wis-

---

[1] Mary Midgley, "Heaven and Earth: An Awkward History" in *Philosophy Now* 34 (2001/2002), 18–21. See also, *The Myths We Live By* (London: Routledge, 2003), 122–127, where Midgley reiterates her arguments.

dom. Natural science is a way of getting acquainted with the living world in all its complexity. The world is an intricate web of relations, of connections; it is also the realm of wonder. Wisdom, therefore, comes into a way of life that can be appropriately informed by science but that goes beyond science and into ethics, art, metaphysics, and, yes, religion. These are all together capable of providing us with a measured sense of our own selves, for instance, our innermost desire to know who we are and what is our place and role in this earthly life.

Mary Midgley has dedicated a career as teacher and writer to the discovery of the sources and reasons for our current attitudes toward nature, other creatures, and ourselves. The attitudes that she seeks to correct are, on the one hand, those that give an overblown view of our human dignities and capacities; this leads into "anthropocentrism," the idea that we humans are at the center of nature or creation. On the other hand, we suffer from an undervalued appraisal of our common roots with every other living thing. We are neither angels nor viruses, neither absolutely altruistic nor irremediably selfish. From the beginning of her public discourse, Midgley has engaged in many battles and confronted bitter controversies trying to drive home what she considers the basic tenets of human nature.

Problems with Evolution

Why does Darwinian evolution cause so much controversy? From its very initiation as a scientific theory by Darwin himself and to the present day, the debates surrounding it seem to be endless. It appears that many people, both from the general public and even some experts, have difficulties in coming to terms with evolution. Given its stormy history in the West, it could also be thought of as a Christian problem or at least a problem that many Christians seem to have with the theory. Many seem to believe that evolutionary theory is helplessly or even necessarily atheistic, as has been time and again implied.[2] This anti-religious aspect alone may account for much of the general public's mistrust of evolution. However, I also believe that some of the difficulties are intimately related with matters of perception or, more properly, with epistemological issues. In any case, as we shall see, these are some of the questions that come to mind when considering the science of evolution.

Some people are still making a fuss out of the fact that it is called a "theory." And, yes, it is a theory, in the scientific understanding and the scientific use of the term. It is a proven and, therefore, reliable theory. The fact remains that there is more than enough evidence for it. Evolution has occurred and, as Darwin suggested, natural selection has been one of the major mechanisms at work alongside variation and adaptation.

Darwinian evolution, despite the controversies around it, and in spite of the number of adjustments and fine-tuning that it has required, stands as a major contribution to science in particular and to the history of ideas in general. Ironically, because of the many debates, either from within or without the evolutionary camp itself, the theory has actually been corrected and expanded, and has therefore survived many a battle. It has grown stronger, not weaker, through time.[3] The fact that many seem to use it as support for many individual ideological agendas does not speak

---

[2] In the USA, the fight over the teaching of evolution in science education as either "theory" or "fact" but not both, together with the insistence of creationists that a version of "intelligent design" theory be taught (or given "equal time" so to speak), never seems to abate.

[3] Support for this assertion and the case for the evidential nature of organic evolution are given by, among others, Stephen R. Palumbi, *The Evolution Explosion: How Humans Cause Rapid Evolutionary Change* (New York: Norton, 2001). Palumbi is professor of biology at Harvard University. With lots of experience in the field, he has collected a good number of samples, in plants, insects, animals, viruses and bacteria, on how

against its nature or pronouncements. The Christian religion has also been used and abused in many ways throughout the centuries, but that does not necessarily invalidate its basic tenets.

A related issue is to ask where Darwin stood on religious matters. In general, he seems to have been ambiguous about religion. In a letter to a Mr. J. Fordyce in 1879, he wrote:

> What my views may be [concerning religion] is a question of no consequence to any but myself. But, as you ask, I may state that my judgment often fluctuates... In my most extreme fluctuations I have never been an Atheist in the sense of denying the existence of a God.

Yet,

> I think that generally (and more and more as I grow older), but not always, that an Agnostic would be the more correct description of my state of mind.[4]

Darwin combined moments where he acknowledged the probability of some versions of a design argument from nature, with moments of deep skepticism about human mental abilities to solve such complex questions. In the first instance, he wrote:

> Another source of conviction in the existence of God, connected with the reason and not with the feelings, impresses me as having much more weight. This follows from the extreme difficulty or rather impossibility of conceiving this immense and wonderful universe, including man with his capacity of looking far backwards and far into futurity, as the result of blind chance or necessity. When thus reflecting I feel compelled to look to a First Cause having an intelli-

---

evolution works and how quick and frequent and therefore, observable to humans, it can actually be.

[4] Francis Darwin, ed., *The Life and Letters of Charles Darwin* (original ed. 1887; reprint ed.; New York: Appleton, 1905), 274.

gent mind in some degree analogous to that of man; and I deserve to be called a Theist.[5]

But then,

> This conclusion was strong in my mind about the time, as far as I can remember, when I wrote the *Origin of Species*; and it is since that time that it has very gradually with many fluctuations become weaker.[6]

At times, Darwin thought that religious belief had played a role, albeit limited, in human evolution. A kind of primitive religious thinking has been in place to make sense of the forces of nature, or in any case, forces beyond human control, as well as of our place in the world.

> If ... we include under the term "religion" the belief in unseen or spiritual agencies, the case is wholly different; for this belief seems to be universal with the less civilized races. Nor is it difficult to comprehend how it arose. As soon as the important faculties of the imagination, wonder, and curiosity, together with some power of reasoning, had become partially developed, and would have vaguely speculated on his own existence.[7]

Darwin's aim was not to try to explain religion away. He thought that religious beliefs, among other possible roles in human evolution, reinforced moral considerations in people.[8] In this way, religion is part of our evolutionary heritage. Religion has also played a role in the emerging properties of consciousness. It has instigated a sense of

---

[5] Nora Barlow, ed., *The Autobiography of Charles Darwin (1809–1882)* (expanded ed. 1958; reissued; New York: Norton, 1993), 92–93.

[6] Barlow, 93.

[7] Charles Darwin, *The Descent of Man, and Selection in Relation to Sex* (2nd ed. 1879; reprint ed.; London: Penguin Books, 2004), 116–117.

[8] Darwin, 118–120. One very helpful presentation on this matter of Darwin's impact on religious thinking is John H. Brooke, *Science and Religion: Some Historical Perspective* (Cambridge: Cambridge UP, 1993), especially 280–282.

community, and the value of communal life. It has been helped by, and it has helped in, the evolution of social instincts. By stating that religion is part of the human evolutionary process, and by insisting on the evolutionary nature of moral values, Darwin was not necessarily arguing for the relativity of religious beliefs. He only intended to put those beliefs in proper perspective from a scientific and naturalistic stand.[9] Nevertheless, his views were taken as a way of deflating both absolute moral values and religious truth.

In general, for Christian thinkers during Darwin's time, evolution was an acceptable explanation of variety in nature. Despite Darwin's denials that chance was for him a proper explanation but rather a sign of our ignorance of all the cause for variation,[10] the perception of many Victorian-era Christians was that the theory insisted precisely on chance elements as a sort of explanation. Even those open to evolutionary views still hoped that chance explanations could be left out of the equation.[11] For Christian liberals, this point was a way to differentiate themselves from the more conservatives ones. For the latter, the very idea of evolution through random variation, without direction or purpose, smelled of godlessness, of a theologically as well as teleologically meaningless universe. This objection has not gone away, and apparently never will, since it touches on the nerve of traditional theological understandings of nature.[12]

---

[9]   Brooke, 281.

[10] See, Charles Darwin, *The Origin of Species* (reprint ed.; New York: The Modern Library, 1998), 172.

[11] Brooke, *Science and Religion*, 283.

[12] Consider, for example, the public declarations of Roman Catholic theologian Christopher Cardinal Schönborn, archbishop of Vienna, a known sympathizer of evolutionary ideas, but now reverting to a kind of "design argument." He has said that "evolution in the sense of common ancestry might be true, but evolution in the neo-Darwinian sense—*an unguided,*

In either case, the idea of chance did not stand a chance with the average Christian person.

Before Darwin, naturalists worked under the assumption that they were helping to create a case in support of the veracity and unity of all creation. They thought that all things on earth were related in terms of their divine origin, design, and overall purpose. A case in point is the work of William Paley, who in his *Natural Theology* of 1804, set out to prove that the study of nature through natural philosophy (science) leads into a deeper understanding of God's activity in the world. Paley was not against science, and aimed this particular volume at a larger educated public beyond the academy. Ironically, others used Paley's arguments against evolutionary theory.[13] Paley's intention, however, was not the commonly held view. There were always objections to, or at least concerns about, the use of rational methods of inquiry into nature and nature's secrets for the purpose of getting to know something of the divine wisdom in creation, in the natural order of things.[14]

From early on, Darwin's theories stimulated new scientific research, especially in terms of the search for human origins—including the oft-quoted "missing links" of evolution. Most remarkably, Darwinian evolution prompted

---

unplanned process of random variation and natural selection—is not." And, "the evolution of living beings, of which science seeks to determine the stages and to discern the mechanism, presents *an internal finality* which arouses admiration" (emphases in italics are mine); from *The New York Times*, Thursday, July 7, 2005, opinion page.

[13] See Keith Thomson, *Before Darwin: Reconciling God and Nature* (New Haven: Yale UP, 2005), 6–7.

[14] For a richer view of the disputes among naturalists who were believers, and the nuances to the diversity of positions that they assumed, see Thomson, especially 45–58, for a good summary of the developing arguments and what he calls the "problems at home."

some to elaborate new religious understandings of nature, for instance, in the development of modern theologies, as well as in social and political theories of almost every kind.[15] It motivated, on the one hand, egalitarian ideas about humans, animals and nature as a whole. On the other hand, it contributed a renewed vision of hierarchical life, from beast to human, or, as the saying goes, from "monad to man," also from primitive to modern humans.

All of these raise the question, as John Brooke accurately poses it: Did the prestige given by Darwinian evolution influence or shape the prestige given to social, economic, and political theories that claimed the appellative "Darwinian"? Or was it the other way around? In almost every European country during the nineteenth century, many intellectuals saw Darwinism as the triumph of "scientific naturalism" and, therefore, of atheistic materialism, especially by those who were antagonistic to Christianity. Moreover, this view also somewhat influenced popular opinion, which came to perceive incompatibility between Darwinism and religious teaching.

Evolutionary naturalism was then embraced, not merely as science, but also as a philosophy, that is, a metaphysical understanding of reality, or even like a religion. It became a welcome attempt by many who looked to find a substitute to a doctrine of creation and, finally, to dispense with a creator altogether, and the Creator's "church," for that matter.

Victorian Tales on Evolution

During the Victorian Era in England, however, it seems that accommodation between some kind of naturalism and theism was more often the case than not. Among the generally

---

[15] Suffice to mention the ideologies of Social Darwinism, fascism, and Nazism. They all used reference to the struggle for life and the survival of the fittest as inevitably acting in the developments of the human society, commerce, and even the state.

educated public, attempts to bring together traditional religion and evolutionary thinking became rather commonplace. Granted, the metaphors of conflict and war applied to the relationship between science and religion were used,[16] but mostly among interest groups, such as self-professed materialists and atheists. According to historian Edward Larson, "just as some people instinctively rejected the idea of human evolution, others embraced it for reasons that had little to do with science. Materialists, atheists, and radical secularists had long displayed a certain fondness for evolutionary theories, such as Lamarckism—anything to dispense with God."[17] In addition, there were those who saw themselves as defenders of traditional ideas and societal order, where institutionalized religion played an important role—a legitimating role, that is.

However, it was also the case that forms of theistic evolution came to the forefront of debates very early after the publication of Darwin's seminal work. The proponents of the latter view regarded theism as a necessary element for a complete explanation of humans and the material world.[18] Early on, the core of theistic evolutionism was to regard God as First Cause of all that is, and as the one who guaranteed order, especially through the so-called "laws of nature." The language of "cause" and "order," or "design," was seen as complementary, rather than in competition, with the new science of nature.

[16] The classic examples, because of their historical impact, are those of John William Draper, *History of the Conflict between Religion and Science* (New York: Appleton, 1874), and Andrew Dickson White, *History of the Warfare of Science and Theology in Christendom* (2 vols.; London: Macmillan, 1896).

[17] Edward J. Larson, *Evolution: The Remarkable History of a Scientific Theory* (New York: The Modern Library, 2004), 134–135.

[18] For a portrait of the complex Victorian attitudes to this issue, see Martin Fichman, *Evolutionary Theory and Victorian Culture* (Amherst, NY: Humanity Books, 2002), especially 169–178.

The reaction from interest groups against this kind of conception, which assigned God a somewhat active and stabilizing role, gathered strength around a group of Victorian intellectuals who began to call themselves "agnostic" during the 1870s and 1880s. Prominent among them was Thomas Huxley, who is credited with the coining of the term (agnostic) in 1869. Huxley used "agnosticism" to denote, first and foremost, his personal views on the question of the possibility of metaphysical knowledge, especially when it came to things divine. He much preferred to assert human inability to solve, by strictly rational argumentation, theistic or theological matters.[19] Huxley and others, in the tradition of that other English philosopher Francis Bacon, believed that reason ought to be followed up to where it leads, and no more. In order to answer non-empirical questions—that is, assuming that science only and exclusively deals with empirical phenomena—for example, about miracles, the existence of God, and the afterlife, we would need to go beyond a legitimate use of scientific reason.

Huxley et al. were convinced that non-empirical investigation demands a kind of knowledge, hence the Greek word "gnosis," not easily available or even really attainable by the tools provided by natural science. It has been argued that Huxley's tendency was to equate rational with scientific, and to further distinguish the latter from the religious language and practice.[20] In any case, many after Huxley's fashion began to call themselves agnostic, too. They formed

---

[19] Fichman, 175.
[20] Fichman, 177.

a movement, which increased dramatically in numbers during the last decades of nineteenth century England.[21]

A Contemporary Tale: Sociobiology

As well as being heirs to the Victorian attitudes described above, contemporary Western societies have a more recent account of the supposed or perceived anti-theistic consequences of evolutionary theory in the doctrine of sociobiology, especially as propounded by Harvard entomologist Edward Wilson. According to Wilson, sociobiology is an "explicitly hybrid discipline that incorporates knowledge from ethology (the naturalistic study of whole patterns of behavior), ecology (the study of the relationships of organisms to their environment), and genetics in order to derive *general principles* concerning the biological properties of *entire societies.*"[22] In this tale, evolution is looked upon as nothing short of a unifying explanatory principle of everything natural—mind and religion included.

We may ask at this point: Is natural selection, that core evolutionary mechanism, working through genetic chance and environmental necessity, truly an alternative to God in the context of the created order? Wilson proposes that evolution, as a product of scientific materialism and naturalism, is a sufficient explanation for all physical and mental phenomena.[23] From this perspective, he proclaims that religious beliefs themselves can be considered and explained as mechanisms of survival. "Religions, like other human institutions, evolve so as to enhance the persistence and influence of their practi-

---

[21] Another helpful source on these issues is the work of A. N. Wilson, *God's Funeral* (New York: Norton, 1999), especially 198–199.

[22] Edward O. Wilson, *On Human Nature* (25th anniversary ed.; Cambridge, Mass.: Harvard, 2004), 16.

[23] Wilson, 10.

tioners."[24] Therefore, there is no need to appeal to any other causes outside the realm of nature, the material substrate of all reality. Natural explanation suffices.

To make his case, Wilson needs to tackle first what he calls the dilemma of naturalism: whether it can actually speak of purpose or goal at all. Evolutionary explanation is not teleological, in the philosophical sense. The assumption is that evolution is driven by chance, namely by blind forces. Any talk of the existence of goals external to nature and natural processes, is not scientifically warranted or even appropriate.[25] In the case of humanity, Wilson was one of the earliest proponents of a gene-driven approach to human nature. In his seminal book, *Sociobiology* (1975), Wilson spoke about the "paramount importance of spreading genes" as a basic human instinct.[26] He also situated the "development of human 'noblest' traits" (like team play, altruism, bravery) as "genetic products of warfare," as the result of previous evolution and adaptations.[27] This was another way in which he intended to demonstrate that there is no goal or purpose external to our biological nature. Whether we are dealing with human emotional response or with human rational faculties, these are all thought to have evolved naturally too. That is, the development of thought (ideas) as well as the capacity for ethical behavior, have been programmed in our brains through a long and gradual process of selection—by trial and error, so to speak.

Wilson is never shy about his newly found faith. Sociobiology has been proposed and defended as a correc-

---

[24] Wilson, 3.

[25] Wilson, 2.

[26] See Edward O. Wilson, *Sociobiology: The New Synthesis* (25th anniversary ed.; Cambridge, Mass.: Belknap/Harvard UP, 2000), 572. He applied his thesis to the human experience only in the last chapter of his book.

[27] Wilson, 573.

tive to the traditional anthropocentrism of religion and the social sciences.[28] The human being is not the peak of creation. Human beings are not that special in comparison with many other life forms. According to Wilson, there is enough evidence of our closeness to non-human animals, especially to other primates, such as the great apes. Particularly in comparison with other primates' genetic configuration and social evolution, we are able to attest to the genetic foundation of human nature and social behavior. Science therefore is able to investigate the evolution of human biology, the confection of the nervous and sensory tissues and systems and hence it can analyze the impulses to activities like art as well as religion.

> I am suggesting a modification of scientific humanism through the recognition that the mental processes of religious belief — consecration of personal and group identity, attention to charismatic leaders, mythopoeism, and others — represent programmed predispositions whose self-sufficient components were incorporated into the natural apparatus of the brain by thousands of generations of genetic evolution. As such they are powerful, ineradicable, and at the center of human social existence... I suggest further that scientific materialism must accommodate them on two levels: as a scientific puzzle of great complexity and interest, and as a source of energies that can be shifted in new directions when scientific materialism itself is accepted as the more powerful mythology.[29]

For Wilson, science is not only called in to explain certain natural systems. It can reconstruct the complexity of the natural world, including the human experience, and also provide a synthesis of understanding what is to be human.

All that said, Wilson acknowledges that, when it comes to the understanding of social evolution, there are clear differ-

---

[28] Wilson, *On Human Nature*, 13.
[29] Wilson, 206–207.

ences between humans and other animals. For one, those differences seem to be mostly conditioned by learning and culture. But that is not quite all. Cultural evolution seems to be basically Lamarckian rather than Darwinian. Still, it is prone to a natural explanation. If we only had all the required evidence and understanding of complex cultural dynamics at hand, we would then be able to explain more. The vision would still be materialistic and naturalistic all the way through.

Wilson is hopeful that through some control over our own biology—made possible nowadays through genetic manipulation, although with inevitable biological restrictions—genetic research will open the future for us. Can we or should we chart our future through biological knowledge? Here is Wilson's scientific (or rather scientistic) manifesto:

> Self-knowledge will reveal the elements of biological human nature from which modern social life proliferated in all its strange forms. It will help to distinguish safe from dangerous future courses of action with greater precision. We can hope to decide more judiciously which of the elements of human nature to cultivate and which to subvert, which to take open pleasure with and which to handle with care. We will not, however, eliminate the hard biological substructure until such time, many years from now, when our descendants may learn to change the genes themselves.[30]

This is Wilson's faith: science, especially evolutionary biology, is better suited than other discipline or area of knowledge for the work of building a future for ourselves where not only the practical considerations of daily life but also questions of meaning can be equally handled.

Following closely on Wilson's trail, Richard Dawkins proposes to use science in order to answer, albeit partially, life's big wonders. But here is a paradox: to the question of

---

[30] Wilson, 96.

ultimate meaning, science asserts that there is none. If anything, natural selection creates the illusion of purpose or design. Nature does not care about us in any special way. Nature seems to be rather indifferent to pain and suffering. Yet questions of meaning and purpose in life do matter to us. Therefore, science cannot turn a blind eye to these. However, Dawkins clarifies that not all "why" questions are the same. Some of those questions are sensical, as when we ask about the utility function of a natural or biological system—what are the eyes for, etc.—while others are not—for example, does God will it so, etc.[31] In the last analysis, according to Dawkins, life's utility function is DNA survival, which is maximized in bodies like ours. In our reality as vehicles of DNA and, therefore, as carriers of genes, we can say that nature "cares" for our survival, at least as long as our genes are passed on. For Dawkins, evolution makes clear that a world of meaningless tragedies, and equally of meaningless good fortune, is all there is. The universe as such is neither good nor bad. It just is. The universe cannot be said either to mind us or to work for us. Truly, it has no intentions.[32]

It is true however that in subsequent statements, for example, in his book *Unweaving the Rainbow,* Dawkins tried to tame somewhat his radical assumptions about the basic purposelessness of the world. There he argues that by observing the complexity of the world through science, we detect that the world is not just beautiful, it is so also because it is true.[33] Science gives us a true picture of what

---

[31] Richard Dawkins, *River Out of Eden: A Darwinian View of Life* (New York: Basic Books, 1995), 128–129

[32] Dawkins, 132–133.

[33] Richard Dawkins, *Unweaving the Rainbow: Science, Delusion and the Appetite for Wonder* (Boston, Mass.: Houghton Mifflin, 1998), x–xi, and 63–64. However, Dawkins also tends to dismiss the idea that such an emotion may be akin to religious experience.

the world is and how it works. However, I do not think that he has properly answered to his critics yet.

If the kind of biological nihilism that Dawkins expresses in a few of his books was what Darwin had in mind when he wrote the *Origin of Species*, it is hard to see how could there be, as Darwin himself concluded in his book, that "there is grandeur in this way of life." I cannot think that even a "modified grandeur" of the kind that Stephen Gould has proposed could work here. A world deflated of all meaning does not look like a universe (in the sense of "cosmos") at all. As John Horgan, writing about Dawkins' views, says: "He leaves no room for mystery, meaning, or purpose—or for great scientific revelations beyond the one that Darwin himself gave us."[34] I would add that Dawkins' interpretation of Darwin with regard to the question of meaning is rather narrowed, to say the least.

But Is It Science?

This is Mary Midgley's question to both of them, Wilson and Dawkins. For Midgley, what we get from Wilson and Dawkins are two views of what a particular scientist wants his or her science to be or be able to do. However, their call to science seems to go beyond what is normally expected of scientific disciplines and research to provide for us. It is one thing to supply explanations for particular mechanisms and how those mechanisms play out in the larger picture that we all work so hard at putting together and making sense out of it. But it is another thing to ask science to provide a metaphysic that may serve as foundation for all of our beliefs. Science cannot claim to be a kind of meta-language that is called to judge the validity of all other expressions of what the world is.

---

[34] John Horgan, *The End of Science: Facing the Limits of Knowledge in the Twilight of the Scientific Age* (Reading, Mass.: Addison-Wesley, 1996), 119.

Sociobiology is a false light because it is "reductive" in the sense of ruling out other enquiries, of imposing its own chosen model as the only norm. But, far more serious than this negative drawback, it is also, like many such reductive disciplines, engaged on its own monstrous enterprise of illicit inflation. To balance the austere renunciation of religious ideas and of a normal view of human standing in the biosphere, which Wilson and Dawkins denounce, they offer us a mystique of power, vicarious indeed but evidently, from the fervent tone which celebrates it, none the less exciting for that.[35]

On the one hand, there is Wilson's worldview, which, as Mary Midgley suggests, it is the kind of scenario that requires a strong faith, not just in human capacities, but also in science's power to promise and create a better future and a better us, in other words, to take control of our own evolution from now on. These are really big dreams and plans for a species that is supposed to be no higher, better or more special than other creatures. On the other hand, we are confronted with Dawkins' views, which could only come from a kind of "metaphysical fatalism"[36] that, by its very nature, falls outside of traditional empirical science.

In contrast to Dawkins' take on this matter, see what Midgley has to say:

> These very varied beliefs, and others equally far from our own, have not prevented people from finding meaning in their lives. Indeed, many have insisted that a full acceptance of the dark perspectives around us is absolutely necessary if one is even to start on an effective search for meaning.[37]

---

[35] Mary Midgley, *Evolution as a Religion: Strange Hopes and Stranger Fears* (revised ed.; London: Routledge, 2002), 154.

[36] Midgley, 150, 152.

[37] Midgley, 107–108.

In Midgley's quote above, she has in mind what she considers the fatalism of such figures as the famed biologist and philosopher of science Jacques Monod, and the physicist Stephen Weinberg, and which is applicable to Dawkins himself. But Midgley takes her argument one step further, claiming that accident and tragedy — the darkness in life — is often the start of the search that sometimes leads to religion itself, and to finding meaning when there was supposedly none to be found. To many, it is the very acceptance of life's limitations as well as its possibilities that is intrinsic to the process of becoming consciously human.

When Evolution Plays Religion

Is the idea of evolution a substitute rather than an antidote against religion or even a form of religion itself, wonders Mary Midgley? This is not an easy question to answer. For a start, it draws strong reactions from opposite parties. The mere suggestion gets a strong negative reaction from those, like Richard Dawkins, who openly mock the very idea that science can be thought of as a kind of religion.[38] On the one hand, Dawkins recycles the argument that religion is based on faith, and science on verifiable evidence. However, he goes beyond this traditional line of argumentation by adding new elements to it. For example, he insists that science is more honest and even moral. "Science... has many of religion's virtues, [but] it has none of its vices." Moreover, "science is free of the main vice of religion, which is faith."

On the other hand, Dawkins still makes science sound like a faith when he states, for example, that "science can

---

[38] See Richard Dawkins, "Is Science a Religion?" *The Humanist: A Magazine of Critical Inquiry and Social Concern* (January/February 1997), also available in *The Humanist Online* <http://www.the humanist.org/humanist/articles/dawkins.html>. This is part of Dawkins' acceptance speech for the honor bestowed on him as the 1996 Humanist of the Year by the American Humanist Association.

offer a vision of life and the universe which, as I've already remarked, for humbling poetic inspiration far outclasses any of the mutually contradictory faiths and disappointingly recent traditions of the world's religions."[39] Among other things, Dawkins keeps shifting gears between a reductive understanding of faith as blind assertion, while at other times using it to describe what he already assumes is a superstitious belief.

Beyond Dawkins, we also get the ravings from those who can only see science as a creed, probably a secular religion. In some cases, the aim is to be able to explain religion, if not to explain religion away. Yet for others the expectation seems to be for science to become a kind of substitute for religion itself. In fact, we could find many examples of each of the above tendencies.[40]

Midgley calls for a different approach. First of all, she is not science bashing. Moreover, she is a convinced evolutionist. Midgley has spent considerable effort in studying the evolutionary origins and basis of human behavior, including our cognitive functions as well as our ethical development. Nevertheless, for Midgley evolution is a powerful myth of origins. We should keep in mind that myth in this context should not be confused with a fable or even less with a lie. A myth is a story, one that seeks to answer the most basic questions about where we come from and for what reason. In this sense, evolutionary science, as it speaks about beginnings, has an important story to tell.

---

[39] Dawkins, 1-2, 6.

[40] For our present purpose alone, it suffices to mention here Edward Wilson, *Consilience: The Unity of Knowledge* (New York: Knopt, 1998), 265. Although he admits that explaining religion and morality is a daunting task for science, he anticipates that eventually science will be able to uncover "the bedrock of the moral and religious sentiments" and the end result will be "the secularization of the human epic and of religion itself" by science.

Midgley believes that it is in the nature of science not to be content only to observe and describe. It seeks to explain, to understand and, therefore, either to fill a vacuum, which is present for lack of explanation in the first place, or to challenge received wisdom with its own.

The above also means that the scientist is never free of motives. Scientists do not collect facts indiscriminately. Together with the facts, imagination plays a role in forming pictures about the world. Facts will almost inevitably lead or form some sort of story or drama.[41] Story telling is an old and very human activity: why would science be free from it or above it? Darwin himself did that: he looked for metaphors, for a language, a story that made sense of the evidence and that could be used to explain the facts. Darwinian evolution has proven to be a resourceful, resilient, and meaningful myth of origins.

At the same time, as a myth of origins,[42] by telling us who and what we are, and by influencing not just our thoughts but our feelings as well, evolution goes beyond its function as a biological theory: it is also a philosophy, a world view, and even, in some instances, a religious perspective.

A number of elements and functions that are traditionally applied to religion have been regrouped under the banner of science, particularly, or so says Midgley, around the concept of evolution.[43] People who express their trust in science by stating, for instance, that it alone is a legitimate intellectual method of human inquiry into truth, are doing a kind of metaphysical thinking normally at home within religious world views. They manifest a strong confidence in

---

[41] Midgley, *Evolution as a Religion*, 1–4.

[42] Midgley has also called evolution the "creation myth of our age;" see, for example, her statements and dialogue in *Discussing Darwin: An Extended Interview with Mary Midgley* (London: Theos, 2009), 15–17.

[43] Midgley, *Evolution as a Religion*, 34.

science's supreme ability to provide answers and lead the way towards human bliss. For Midgley, this is not merely science, but rather faith. No doubt that faith is part of any human endeavor or activity. But the faith that plays a role in many human activities is normally very limited in scope.[44] Thus we say that we have faith in our school system, or in our marriage, or in people to do what is right when it comes to voting in general elections. But today, people who expressed an especially strong faith in science are basically claiming that its method can be applied to many, if not all, areas of the human experience.

In the case of evolutionary science, faith in a superior and inevitable evolution seems to be strong in those whose hopes are placed, for example, in genetic engineering and the enhancement of human faculties.[45] This feels like an offer of salvation, a pie-in-the-sky promise. However, it has been a common charge against religion that it promises fulfillment, deliverance, and consolation by pandering to people and soothing them into believing unwarranted graces and predictions. According to Midgley, it is here that we enter the realm of prophecy. Prophecies, in traditional religious discourse, are patent in dazzling predictions about the future. But so it happens with science.

---

[44] Midgley, 23–24.

[45] Publications on the promise of human enhancement come out almost weekly. From the exponential growth of human capacities to the increment of human intelligence and the extension of human lifespan, all possible through genetic manipulation and brain modification, the prophecies for a new future are abundant. One striking sample is the recent volume by Ray Kurzweil, *The Singularity is Near: When Humans Transcend Biology* (New York: Viking, 2005).

Midgley mentions at least two intimately related examples of this prophetic tendency in science.[46] The first comes in the prospects for the further evolution of human intelligence and the development of the human species. Notions of a "superman" have been proposed several times over. Biologist William Day (b. 1928) has proclaimed that, "he (man) will splinter into types of humans with differing mental faculties that will lead to diversification and separate species. From among these species, Omega man, will emerge either alone, in union with others, or with mechanical amplification to transcend to new dimensions of time and space beyond our comprehension."[47] Moreover, there are high hopes for the future of genetic engineering as well as artificial intelligence. We are often led to believe that we are fast approaching the time in which we will be able to enhance ourselves, increased our multifold capacities by proper engineering of the human genome or maybe the brain.[48] We are also asked to believe that it is both good in itself and good for all. The offer in this case is nothing short of immortality and the crafting of the meaning of life. It becomes a discourse on human destiny and the ways to control it, for our own good, the good of the species.

Neither science nor any other discipline or research can guarantee a future made to our specific requirements. The world, or the "laws of nature" for that matter, is not set to benefit us necessarily. Nothing says that science is either able or that it should even try to deliver on an offer of emotional support or a promise of bliss on earth or "pie in the sky" solutions to our utmost challenges and needs. More-

---

[46] See, Mary Midgley, "The Religion of Evolution," *Darwinism and Divinity: Essays in Evolution and Religious Belief* (ed. John Durant; Oxford: Basil Blackwell, 1985), 156.

[47] William Day, *Genesis on Planet Earth: The Search for Life's Beginning* (2nd ed.; New Haven: Yale UP, 1984), 390-391.

[48] Midgley, *Evolution as a Religion*, 37, 78.

over, if science has been used so often to belittle religious faith, it cannot all of a sudden become a faith itself. Science should not be thought of as a form of salvation. There are certain teachings deemed to be scientific, which tend to conflict with religion. In some cases, and we have mentioned the case of sociobiology, they may even seek to be a substitute for religious doctrine. They aim at being salvific to humanity, in the sense of attempting to fulfill spiritual needs beyond a mere acquisition of knowledge about the physical world or the human predicament. The bottom line is that they are neither scientific nor religious in a proper sense. Midgley addresses this issue with some examples. She argues that,

> Marxism and evolutionism, the two great secular faiths of our day, display all these religious-looking features. They also have, like the great religions and unlike more causal local faiths, large-scale, ambitious systems of thought, designed to articulate, defend and justify their ideas — in short, ideologies... I think that to say that Marxism or evolutionism, or indeed art or science, is serving as a religion, can be a useful way of speaking today... They are, not accidentally but by their very nature, dominant creeds, explicit faiths by which people live and to which they try to convert others. They tend to alter the world.[49]

It is remarkable the way in which Midgley uses a parallel between Marxism and evolutionism. For her they could both be considered as ideological systems, which engage in power struggles for political gain in society. These are both secular faiths, because of their basic intention to occupy the place in our lives that religious commitments have occupied for many peoples around the world. Claiming total acceptance and loyalty has been part of their "mystique" too. Meanwhile, they have been for the most part failed

---

[49] Midgley, 17-18.

attempts at stripping people's minds and hearts from all remnants of the old faiths.

Their systems of thought are dogmatic in character. As overreaching dogmatic beliefs, they exhibit an intrinsic refusal to even consider different views. More critical yet for Midgley is the fact that these two conceptual systems are also forms of Utopia,[50] which see human progress towards a better future as inevitable, if painful, for the time being. This kind of belief falls already in the realm of prophecy. In addition, they show a tendency towards expressing a kind of missionary fervor in spreading and defending their ideas. And, on top of all that, they also convey a sense of the melodramatic in human history.[51]

Dogmatic Dreams, Or the Illusion of Impartiality

What Midgley is pointing to is a kind of dogmatism, epistemological in character, which is not rare in science nor, therefore, in evolutionary biology. It actually thrives in ideologies of many types. It is the dogmatism attached to an insistence on pure or absolute objectivity, the kind of analysis supposed to be completely free of values when assessing matters of fact and evidence. But the fact remains that, "the arguments for our own faiths, including science itself, lie outside science too. If we have the impression that our own faith needs no argument, being simply self-evident, this is merely a dogmatic slumber."[52]

Midgley mentions the famed French biologist Jacques Monod as a case in point, especially his book *Chance and*

---

[50] Midgley, 66.
[51] Midgley, 89.
[52] Midgley, 23.

Necessity.[53] It is Monod's intention to rid biology of any instance of what he called "animism," that is, attributing conscious purpose to the universe as a whole or to any being in producing life and, therefore, us, sentient and intelligent beings. Monod is also critical of any claim to the so-called "laws of nature" to create a sense of necessity or direction or pre-existent order that could explain our being here in this kind of universe.[54] By doing this, Monod then excludes any metaphysical considerations from science. His argument is for the exclusivity of the scientific explanation, untainted by any subjectivism, as the only true form of valid knowledge of the world.

According to physicist and philosopher of science David Bohm (1917–1992), this postulate of objectivity in science is akin to an article of faith in religious traditions.[55] Midgley is in basic agreement with Bohm, but she also adds that the kind of argumentation for objectivism that Bohm criticizes is not good scientific reasoning either. It is rather arbitrary. It pretends to subordinate all other ideas and forms of reasoning to scientific knowledge. She does not think that there is any legitimate reason or cause to do this, and it would not work anyway. This is because such a science would be seeking to understand the universe without a unifying metaphysics that is able to draw on different forms of knowledge and analysis, or as John Haught would argue, one able to "read different levels of reality." That kind of science would be basically a mindless and meaningless collecting of lim-

---

[53] Jacques Monod, *Chance and Necessity: An Essay on the Philosophy of Modern Biology* (trans. Austryn Wainhouse; New York: Knopf, 1971).

[54] Midgley, *Evolution as a Religion*, 89.

[55] See David Bohm, "On the Subjectivity and Objectivity of Knowledge," *Beyond Chance and Necessity: A Critical Inquiry into Professor Jacques Monod's Chance and Necessity* (ed. John Lewis; London: Garnstone, 1974), 126–127.

ited facts about the world.[56] As long as science strives for a comprehensive picture, a picture of the whole, it will inevitably open into metaphysics.[57]

What Midgley and the others are saying is that, in general, and for the intelligibility of their own work, scientists assume the unity and comprehensibility of the world. However, and at the same time, there is a tendency to deny the need for such an assumption. At times, they even deny making such an assumption of necessity. When science gets into that kind of contradiction, the scientist seems to be of a split-mind, and his or her science, in the words of British philosopher of science Nicholas Maxwell, becomes "neurotic."[58] By denying or avoiding or not acknowledging what they do and need to do, scientists have to give all sorts of twisted explanations about their own activities.

At times it seems that scientists have become quite dismissive of other perspectives or areas of human thought and reflection. Many of their intellectual claims are made off-hand, without proper argumentation.[59] Their peculiar beliefs and maybe even morals, as in the case of evolution, can be also said to have become like a religious faith. Its acceptance as well as the strong confidence in its invulnerability is akin to a dogma. It dictates the scientist's beliefs and values.

All that said, it is equally true that science is not a closed system, isolated from other areas of thinking and living. It is, rather, a system that connects with the rest of our thinking. It is not neutral either. Scientists seek to explain, and sometimes, in the midst of the competition of ideas (and ide-

---

[56] John Haught, *Deeper than Darwin: The Prospect of Religion in the Age of Evolution* (Boulder, Colo.: Westview, 2003), 13–25.

[57] Midgley, *Evolution as a Religion*, 121.

[58] See, Nicholas Maxwell, *Is Science Neurotic?* (London: Imperial College, 2004), 1–3, for a thought-provoking argument.

[59] Midgley, *Evolution as a Religion*, 110.

ologies), they seek to explain away whole practices or systems of thought. In the pursuit of knowledge, it is one thing to say that we should keep personal bias from affecting our thoughts and the results of our investigation. But it is another thing to say that we cannot acknowledge or even think about subjective factors at all. The findings of science, especially of the physical sciences, "are supposed to be objective, that is, free of bias."[60]

The key concept here is impartiality. Scientists normally claim full objectivity when dealing with the facts of the world. But impartiality cannot mean the exclusion of any motivations, interpretation of facts, or a requirement to think of subjective states as merely unscientific. That kind of impartiality is rather an illusion. Subjective factors are actually important in the kinds of decision that the scientist has to make concerning the focus, extent, and possible applications of his or her research, among other motives. In any case, acknowledged or not, positively or negatively understood, subjective factors are present in the scientist's work whether we like it or not, or even if we wish them away.

Unfortunately, the argument for objectivity and impartiality has been used against any consideration for the reliability of the religious experience of the world. Religiosity has too often been termed subjective, and faith considered blind, non-rational, and thriving in "mystery," by certain influential scientists and philosophers of science.[61] Midgley

---

[60] Midgley, *The Myths We Live By*, 78.

[61] See, Richard Dawkins, "Viruses of the Mind," in *Dennett and His Critics: Demystifying Mind* (ed. Bo Dahlbom), 20–21. This way of describing faith is very typical with Dawkins, but I could be equally referring to what chemist and philosopher of science Peter Atkins says about faith, that it is a delusion, a combination of ignorance and fear, and baseless. See, for example, his essay "The Limitless Power of Science," in *Nature's Imagination: The Frontiers of Scientific Vision* (ed. John Cornwell; Oxford UP, 1995), 125.

sees at least two mistakes in these assumptions. One mistake is to think that the study of subjective phenomena is the same as being subjective, or being carried away merely by emotions, moods, and feelings.[62] A second mistake is to think that reflecting on the inner experience, or attending to psychological factors, is the only thing that religion is about. Moreover, it is misleading to think that religion is so isolated from reasoning or thought as to make it a mindless activity and non-sense language.[63] The people who make these kinds of mistakes tend to be very selective in the way that they apply reasoning, and where and where not to put their doubts.

Darwin's Selective Doubts

For Midgley, although Darwin acknowledged that there were reasons for belief that were teleological as well as cultural, he still chose for himself to doubt the reliability of any propensity to believe in God. Darwin's was a rather selective doubt, since he did not put his own choice of disbelief or his own confessed agnosticism under the same suspicion.[64] The problem for Darwin was that he convinced himself that belief in God had its roots first in biological evolution and then in cultural demands and usage.[65] His biological assumption would explain for him why the God-idea could last for so long as part of the human consciousness of the world.[66] Darwin thought that belief in God could have come from "the lowly origin of our minds," something which seems to have made him quite uncomfortable. However,

---

[62] Midgley, *The Myths We Live By*, 145.

[63] Midgley, 105.

[64] See Mary Midgley, *Science as Salvation: A Modern Myth and Its Meaning* (London: Routledge, 1994), 101–102.

[65] Barlow, *The Autobiography of Charles Darwin*, 92–93.

[66] Midgley, *Science as Salvation*, 102.

the same "lowly origin" could be argued of many of his other ideas, including his ambiguous faith, or evolution itself, for that matter. But that is not what he did with the latter. Therefore, according to Midgley, in regard to faith and God, Darwin remained selective in his doubts.

Darwin's skepticism probably owes more to ideas about reason, from Enlightenment ideals, than from his own evolutionary theories, as he seems at times to imply. Besides, why assume in the first place that the idea of and belief in God are simple, basic, and old, in the sense of primitive? To think, for instance, that our minds and subsequent ideas have evolved from lowly origins, or from no minds at all, is no reason to doubt of our current capacities for reasoning. Darwin himself did not stop thinking and writing because of the evolutionary character of human faculties.[67] To discard ideas about God or any grand questions because they may have come from far back, is not something that Darwin necessarily does with other concepts, tendencies or practices. Once more, it does bother Midgley that Darwin himself had been so selective in his reason for his doubts about religion. Any grand truth about the world would be beyond us in the same way.

For Midgley, the reverence that Darwin showed for nature and its evolution was truly laudable.[68] It did not have, however, to drive him into skepticism or agnosticism as he suggested in his autobiography.[69] How come that intelligent and otherwise unbiased researchers show so much trouble when it comes to understanding religious ideas and practices? Partly it may result from the very complexities of defining such a multi-layered and pluralistic experience. But there are also those with a certain tendency to analyze and doubt the rationality of other people's ideas

---

[67] Midgley, 105.
[68] Midgley, 106.
[69] Barlow, *The Autobiography of Charles Darwin*, 94.

and not their own. Darwin has not been the only selective skeptic after all.

Evolution Explaining Religion

Can religion be explained or explained away by evolutionary theory? Is religion a survival mechanism? Is it a biological adaptation? Is it the product of environmental and life changes during the development of humans from times long past? Maybe a carry-over from a by-gone era, say, the hunter-gatherer societies of the Pleistocene two million years ago? These are the kind of questions that we may ask when confronted with the question of the origin and meaning of religion from an evolutionary perspective.[70]

The answers, as can be expected, vary from one scholar to another. It all depends on where the scholar stands in regard to his or her analysis. Religion itself then is put under the microscope in order to sustain a Darwinian account. A case in point is philosopher Pascal Boyer on the evolutionary origins of religious experience.[71] Boyer believes that cultural causes are secondary to biological ones when it comes to religion. Therefore, he wants to dispense with ideas about God and revelation when it comes to explaining religion and why people are religious. By insisting that the persistence of religion is a consequence of the way that natural selection designed our brains, he also aims to demonstrate the diminishing efficacy of religious ideas and practices when it comes to the question of survival today. According to Boyer, the problem becomes more acute when our old habits and fears may be helping to conceal from us the pass-

---

[70] See, for example, John Haught, "The Darwinian Universe: Isn't there Room for God?" *Commonweal* 129/2 (2002), 12–18.

[71] Pascal Boyer, *Religion Explained: The Evolutionary Origins of Religious Thought* (New York: Basic Books, 2001).

ing utility of religion as well as the truth of its malfunctioning in the modern world.[72] Sometimes evolutionary theory is even used to make a case for science and against faith. We have already named, and even quoted from, the likes of Richard Dawkins and Edward Wilson. In a reductive approach, thinkers like them still believe that by getting rid of ideas about God and the sacred, they could explain, or explain away, religion. Others are only looking to understand something peculiar about human nature: our propensity for religious thinking and ritual, or the role of religious emotions in helping us to construct a world of meaning for ourselves as well as for others.[73] This is done in disregard to the question of truth in favor of a concern with practical living. Moreover, some Darwinian anthropologists think of this human self-deception or evasion of truth, of which they accuse religion, as a kind of adaptive mechanism, one that can make us think that we are being helped, so to speak, and that life is worth living anyway.[74]

It has become rather common to assert the adaptive value of a culture or a cultural trait in helping people to adjust to a specific environment. Thus, it can also be assumed that religion, being such a common and almost universal human experience as well as so intimately related to culture and society, contributes somehow to human growth and survival — and this despite all of its difficult manifestations and

---

[72] Boyer, 22; in this way, Boyer seems to personalize 'natural selection' by attributing to it plan and purpose.

[73] A more recent attempt to explain religion by strictly naturalistic means is that of the philosopher Daniel Dennett, *Breaking the Spell: Religion as a Natural Phenomenon* (New York: Viking, 2006). Dennett intends to demonstrate that religion is itself biologically evolved, and that even its cultural aspects have an evolutionary and, therefore, naturalistic explanation — the theory of 'memes', for instance.

[74] See Haught, "The Darwinian Universe," 13.

history. If nothing else, religion does have a role in the formation of group identity. But, in truth, it can also be a means of passing on information about a group's precedence, development, and commitments. Religion is a transmitter of traditions too.[75]

However, the question of whether specific religious practices or ideas, or any other religious phenomena, have been naturally selected by their ability to increase the chances of group or individual survival, is a more difficult question to answer correctly. One way to answer it is to consider practices regarding purity and hygiene, family and reproduction, piety and work, all of them contributing somewhat to the reproductive success of the individuals that follow them.[76]

Natural selection is sometimes believed to operate among groups, in addition to individuals within groups. This is an assumption made by a few biologists as well as anthropologists, despite the arguments concerning the specifics of the mechanisms involved.[77] In the case of religion, some would argue that adaptive explanations of this phenomenon include those working at the group level as well as at the individual level. Within this view, there are those who see religion as a cultural parasite or virus of the mind that can and has spread through individuals and groups, thus ensuring their own survival and reproduction.[78]

Another strategy would be to take the route of a non-adaptive explanation. The latter seems to lead in two

---

[75] A helpful discussion on these issues is part of Bernard J. Verkamp, *The Evolution of Religion: A Re-Examination* (Scranton, Penna.: University of Scranton, 1995), 125–151.

[76] Verkamp, 128.

[77] See Scott Atran, *In Gods We Trust: The Evolutionary Landscape of Religion* (New York: Oxford UP, 2002), 264.

[78] We have cited Richard Dawkins, "Viruses of the Mind," 18–19. See also, Susan Blackmore, *The Meme Machine* (Oxford: Oxford UP, 2000), 201–202.

distinctive directions: on the one hand, to insist that religion is not an adaptation and that it has no evolutionary function as such. For one, there is no entity called 'religion.' On the other hand, it would be said that a religion or religious practice in question was first adapted to other, past environments, but it has now become mal-adaptive to present circumstances.[79] Finally, some like to argue that religion is but another by-product, albeit a significant one, of the combination of some genetic and cultural evolution. It is like the development of language or writing, which came only after significant growth of the human brain. Religion itself would be then considered one of those "spandrels" of which Stephen Gould spoke so much.[80]

All of the positions mentioned above assume, nevertheless, a close interrelation between biological and cultural evolution.[81] That relationship is also sometimes termed a debate between the biologically-natural and the socially-nurturing aspects of human development. In any case, it is important not to assume that cultural adaptation is merely an extension of the biological kind.[82] To assume that would be like looking for an adaptive solution to any identifiable cultural trait. As a matter of fact, cultural practices and

---

[79] Atran, *In Gods We Trust*, 265. The current maladaptive nature of religious ideas and practices is a point often made by the supporters of the 'memetic' understanding of religion's original functionality, such as Daniel Dennett, Susan Balckmore, and Richard`Dawkins, to name a few.

[80] For the classic statement of Gould's position, see Stephen Gould and Richard Lewontin, "The Spandrels of San Marco and the Panglossian Paradigm: A Critique of the Adaptationist Programme." *Proceedings of the Royal Society of London. Series B: Biological Sciences* 205 (1979), 581–598.

[81] For a helpful analysis of many of these positions, see David Sloan Wilson, *Darwin's Cathedral: Evolution, Religion, and the Nature of Society* (Chicago: The University of Chicago, 2002), the whole of chapter one, 1–46, but especially, 44–45.

[82] Verkamp, *The Evolution of Religion*, 138.

development are very complex phenomena. They defy for the most part single explanatory principles or simple solutions. Culture itself is a tangled web of relations, where it becomes very difficult to say what practices generated which other ones. However, the question of adaptation brings us to the one about cultural evolution: what it means, and whether it can be thought of as following a parallel track to that of human biology.

Midgley has dealt with the issue of the relationship between biological and cultural evolution. She understands the need for finding correlations between the two.[83] Ideally, there would be one principle that could explain both biological and cultural developments. A kind of scientific rationality seems to demand a simple explanation. That kind of rationality, also referred to sometimes as "Ockam's Razor," says that we should postulate as few principles as possible when explaining natural phenomena, including human evolution. Therefore, finding an analogy between these two kinds of evolutionary processes, or two kinds of selection, might shed clarity on the whole issue of the development of civilizations, and also the history of ideas, among other possibilities. It is also worth thinking about the continuity between human life and the rest of nature, or the created order. Midgley is very much in favor of the kind of research that can bring us humans closer to other species.[84]

However, Midgley also warns that such analogies could only take us so far. Despite elements of continuity, there is no indication of universal laws that would equally apply to the realms of biology and history.

> Continuity is not identity. The unity of science is real, but it can't be stated so simply. It requires compatibility between the results of different enquiries, but not formal unity.

---

[83] Mary Midgley, *Biological and Cultural Evolution* (Kent, UK: The Institute for Cultural Research, 1984), 6.

[84] Midgley, 7.

Forms of explanation are bound to vary, because different sorts of subject matter present different problems to our understanding."[85] In addition, there is the question about the existence, even in science, of one single form of explanation. There is not a single form of thought, like one "scientific method," to which all explanations can be reduced. As scientist and philosopher John Ziman (1925–2005) has forcefully argued, science is a social network of research and practices and, as a network, is not regulated by any single rule, principle, method or agency. Science is a complex whole. The very idea of what science is does not originate or belong with any one person or group alone.[86] If science is able to contribute, and in some cases significantly, beyond those areas traditionally deemed to be its own turf, it is precisely because of a plurality of tools and approaches that it uses in actual practice.

On the issue of finding meaningful parallels or patterns between cultural and biological evolution, there is the additional complication that similar methods could mean different things in different contexts, and also often produce different results. Again, take for example what Midgley has to say concerning the idea of "selection:"

> Cultural selection is selection in a directly literal sense; there always are actual people who literally choose, select, and prefer one thing rather than another. However confusedly they act, they always have their motives, their reasons at some level. But in biological selection there are no such people and no extraneous purposes for which the results must be "fit." The analogy between the two kinds of selection can therefore never be complete. It is often illuminating, but only passing — it must never be pressed. To

---

[85] Midgley, 8.
[86] See John Ziman, *Real Science: What It Is, and What It Means* (Cambridge: Cambridge UP, 2000), 1. Reading Ziman's work on this and related issues has been recommended by Midgley herself.

insist on tackling human history by the methods of biology seems like a perverse importation of ignorance.[87]

Now, if what we bring is the question of "purpose" upfront, the picture changes.

> Reference to... purposes and knowledge of them is not a legitimate tool in biology, but it is a perfectly legitimate tool in history; in fact, it is an essential one.[88]

The bottom line for Midgley is that there is no universal law, and no idea of necessity, that could be extrapolated from the realm of biology into history or vice versa. None has a single direction either upward or lineal. In addition, among humans, individual selection has given way for the most part to cultural change. "Culture is now the dominant adaptive mechanism of our species and has its modes of changing."[89] Finally, "the time scale, both of evolution and of cultural change, is far too great for us to have the slightest hope of identifying their general direction from within."[90] Even if it had a temporary direction, we would not be able to know it, properly speaking.

What is Midgley's basic point? She argues that we have a nature as well as a history, and that to think that they both could be subject to identical methods and analysis, not to say similar or parallel tracks, strikes her not just as naïve but also as a way to avoid doing some thinking. "What worries me is the hasty use of certain patterns that have been found useful in biology to explain human affairs where they have only a somewhat artificial application, at the expense of the directly relevant study of human motives."[91] Besides, change in human societies is now almost entirely due to

---

[87] Midgley, *Biological and Cultural* Evolution, 6–7.

[88] Midgley, 6.

[89] Midgley, 13.

[90] Midgley, 14.

[91] Midgley, *The Myths We Live By* (London: Routledge, 2003), 87.

cultural factors, rather than genetic or biological.[92] Unfortunately, Migdley does not expand on this latter point. In any case, it seems plausible to assert that human evolution and civilization have come to the point in which our doings, the way we impact and therefore shape our environment, has made of cultural change and development a major determinant of what lies in the future for humankind.

The Need to Come to Terms with Evolution

A growing number of biologists have been insisting that we have a place in this world of ours, the earth. We are not aliens here. The earth is our place. It is nourishing and, therefore, sustains us. We do not need to look to the heavens to find meaning. Life on earth is embedded with meaning. Knowledge of who and what we are, as well as our relations to every other living being, can help in the task of building meaning, although it cannot do it alone. Again, Midgley:

> If we drop our ambition to figure as central actors on the cosmic stage, and return to our own field of terrestrial life, we shall find no shortage of purposes, many of them very demanding. We shall probably see the problem of meaning as chiefly one of harmonizing these purposes, and of understanding them well enough not to waste our lives on shadows. Among those purposes, those concerned with knowledge will indeed appear as important, and the sciences, physical and otherwise, as very serious fields of human endeavour.[93]

This is Midgley's way of saying that neither the challenge of complexity itself nor the ambiguity of what being terrestrial might mean, should deter us from careful attention to our place and possible roles in the web of relations that is living in the biosphere.

---

[92] Midgley, *Evolution as a Religion*, 80.
[93] Midgley, 109.

It is this way of thinking about life on earth that I have found particularly helpful in Mary Midgley's philosophy. As a civilization, we seem to have lost a sense of nature and the natural as truly enriching of life in all its dimensions. There has been a loss of the enchantment of what to live means. Some have even argued that Christianity itself has been averse for the most part to considering nature favorably as home, as a nurturing experience, as garden and not just the context of our tears and sweat. Midgley herself seems to believe that Christianity has been more counter-productive than visionary when it comes to a correct and deep appreciation of nature.[94]

However, not everybody agrees with Midgley on this point. Archbishop of Canterbury Rowan Williams (b. 1950) has clarified that the Christian tradition does not have a uniformly bleak record on ecological matters.[95] He has argued that from within Christianity, as with other religious traditions, the earth has also been considered a form of communication or a "system of meanings" that tell us to be attuned to God, while at the same time warning us not to impose our own meanings on the world.[96] We cannot do with creation as we please. Also, Anglican theologian Alister McGrath (b. 1953) questions the idea that Christianity is part of the ecological problem. At least in principle, Christianity has taught that the earth is God's since God is its creator. And, "if nature [is] God's, humanity [is] in no position to mess

---

[94] See Mary Midgley et al., "Environment and Humanity: Friends or Foes?" in *Conversations with the Archbishop* at St. Paul's Cathedral, London, on September 21, 2004; see transcript, 2.

[95] See Williams' comments in this regard in Midgley et al., 5. Williams is the 104th Archbishop of Canterbury, of the Church of England. He is an accomplished theologian, historian and poet, with many published works.

[96] Midgley et al., 5–6.

around with it."[97] In the steps of Williams, McGrath believes that Christianity and other religious traditions have the resources for a rich and meaningful conception of nature and of nature's value. However, in line with Midgley,[98] who claims that modern science is responsible for introducing the image of the world as machinery and inviting us to treat nature as clockwork that can be tinkered with, McGrath sees the roots of the ecological crisis in the emergence of human autonomy and the idea of nature as a mechanism subordinated to human control and advantage.[99]

Truly, evolutionary theory can actually enrich a Christian understanding of the world. It can be the proper background for new views on divine action in nature. Evolution points toward a fuller picture of the evolving world, in which elements of beauty as well as tragedy are part of the picture. Theologian John Haught says that many theologians' obsession with order and design makes it more difficult to get a deeper sense of both nature's own complexity and God's intimate participation in natural history.[100] Moreover, Haught argues that Darwin has helped to rethink our ideas about God, by giving us a more concrete, down-to-earth approach to matters religious and theological—a view more in accord with an "incarnational stance" in Christian thought.[101]

From its inception, there was support for the Darwinian understanding of natural evolution among the clergy and

---

[97] See Alister McGrath, *The Reenchantment of Nature: The Denial of Religion and the Ecological Crisis* (New York: Doubleday, 2002), xiv.

[98] See Mary Midgley, "The Supernatural Engineer," *The Essential Mary Midgley* (London: Routledge, 2005), 284-285.

[99] McGrath, *The Reenchantment of Nature*, xvii.

[100] See John Haught, *God After Darwin: A Theology of Evolution* (Boulder, Colo.: Westview, 1995), 5.

[101] Haught, 45-56.

religious thinkers of Darwin's day. Darwin was more popular in theological and ecclesiastical circles in the nineteenth century than is normally admitted.[102] Ironically, the scientific reaction was more skeptical at the beginning than usually depicted. This is to show that questions about God were not really central to early debates.[103] The main question was rather the place of the human in evolution. It has always been a sensitive issue. It is a question about our origins. It is also a question about human dignity, about our place and role in nature. No doubt the questions, at least for some people, linger.

I am convinced that given the nature of evolutionary theory and how much it is able to explain about our world, it will be as well for the religious thinker to come to terms with it, or at least deal with the possibility of it, and find his or her own way through the evidence. However, it may require a few revisions in thought. For example, traditional theistic conceptions of the divine, especially in his or her relation to the created order, will suffer some modifications. God and creation have a long history. God's deep involvement does not preclude some sense of the messiness of an evolving world.[104] Creation is not necessarily a closed system, but an open one: it is not yet all that it could be or become. We can soundly say then that the universe is still unfinished. The future of creation is therefore intimately related to and dependent on God's own future. Moreover, the world is imperfect and, therefore, suffers from its own imperfec-

---

[102] See Arthur Peacocke, "Welcoming the 'Disguised Friend': A Positive Theological Appraisal of Biological Evolution," *An Evolving Dialogue: Theological and Scientific Perspectives on Evolution* (ed. James B. Miller; Harrisburg, Penna., 2001), 371.

[103] See Mary Midgley, "Darwinism and Ethics," *Medicine and Moral Reasoning* (ed. K. W. M. Fulford et al.; Cambridge: Cambridge UP, 1994), 6.

[104] Haught, *God After Darwin*, 37.

tions, including accident, suffering, and evil.[105] From an evolutionary perspective, we see more possibilities in dealing realistically with these difficult questions.

Open-minded Christians and other religious thinkers have seen advantages in a theology informed by evolutionary theory, not just focusing on problems or pitfalls.[106] It could help to change the views on God's relation to the world — which can be seen as an ongoing creative process. It can also open new venues of dialogue about "divine transcendence" (and the autonomy of nature's laws) and "divine immanence" (as in the case of emergent novelty). It may lead, at least in the case of Christian theology, from a renewed sense of our place in the world to a richer understanding of God's "real presence" in the world.[107]

This is not an attempt to deny the difficulties, theoretical or practical in nature, with assuming an evolutionary epistemology. The realities of struggle, destruction, and evil remain as challenging as ever for any kind of epistemology, religious, evolutionary, or otherwise. However, this is not exclusively a problem for a theology of evolution or any

---

[105] Haught, 38.

[106] I refer here to believers who have an interest in remaining "conversant" with the surrounding culture. I actually see it as an important part of the heritage of the Christian tradition, from early times (the so-called Patristic era, for example) to this day, still there for anyone who is not hesitant or afraid at being challenged in the "comfortable zones", even in his or her beliefs. This is where I see myself. It is not intended as a "put-down" to anyone.

[107] As a Lutheran theologian, this is an important issue for me. Lutheran theology is emphatic on the nature of God's "real presence," not just in the liturgy of the Church and the sacraments of Baptism and Communion, but also in communities, concrete acts of mercy, the "edifying conversation among believers", and furthermore in creation and the natural order — not just as "order" but as "reality."

form of evolutionary theism. This is a problem for any kind of theism.

Darwin's insights and solutions are still partial answers to the kind of challenges that theologians and religious philosophers confront.[108] For example, Darwinian evolution cannot answer questions of the purpose or goal of life in the universe, or of the final predicament of human evolution for or by itself. It cannot say anything about God's being, properly speaking, but it can provoke a new thinking about the meaning of God's creative activity in terms of God's self-giving and ultimate self-sacrificing love in allowing for an evolving and somewhat self-creating universe.[109] Nothing really stops religious thinkers from making Darwin's insights their own. Darwin's gift is a new openness to everything earthly and natural. The idea that being scientific about the world, and taking evolution seriously, is necessarily irreligious is not just naïve but a confusing one.[110]

However, the greatest challenge will always be for any theology, scientific or otherwise, or for any religious philosophy, that wants to keep some concept of benevolent or intelligent design in nature. The latter is not my approach or contention. My aim has been to make clear how evolutionary thinking can actually help in elaborating a religious and "down-to-earth" conception of the world.

---

[108] Brooke, *Science and Religion*, 316.

[109] Haught, *God After Darwin*, 53–55; what Haught proposes is basically a "kenotic" view of divine being and agency, by another name. This modifier ("kenotic") comes from the Greek word *kenoun*, "to empty" or "to abandon," as used in Phillipians 2:7 in the New Testament, where it is applied to the (self-)sacrificing Christ himself, and from which we get the theological term *kenosis*.

[110] Midgley, *Science as Salvation*, 12.

Chapter Five

Midgley's Approach: Thinking from Below

Midgley's Empiricism

In order to understand Mary Midgley's own contributions, as well as the arguments of this book, concerning an epistemology informed by evolutionary theory, we need first to learn the basics of what I like to call her "moderate empiricism." For one, Midgley believes that the traditional philosophical dispute between rationalism and empiricism is not justified. For her, the distinction between these two epistemologies is merely of emphasis. Knowledge depends on both, reason and experience.[1]

Nevertheless, Midgley stands in the tradition of British philosophical empiricism. This is something that she herself has acknowledged, while also insisting on the need to understand empiricism as a rich view of the world and not as a reductive or simplistic one. In Midgley's definition, "the general spirit informing empiricism is simply an emphasis on experience as against theory, a call to attend to the shape of life as it actually hits us, rather than relying on a formalized account of it drawn from abstract schemes."[2]

---

[1] See Mary Midgley, *Heart and Mind: The Varieties of Moral Experience* (revised ed.; London: Routledge, 2003), 6.

[2] Mary Midgley, *Wisdom, Information and Wonder: What Is Knowledge for?* (reprint ed.; London: Routledge, 1995), 200.

Thus, she is committed to empiricism as an assertion of the primacy of experience over theoretical principles in forming our knowledge.[3]

Midgley criticizes the epistemological "dead end" to which David Hume (1711–1776) brought all questions about the physics of the world and the psychology of the human experience. "Hume ... moved steadily away from simply using the realities of experience ... and towards a concentration on constructing a rival metaphysic that could defeat the rationalistic ones."[4] Unfortunately, it is contingency—normally defined as the possibility for an event of not having to occur[5]—rather than the reliability of the experience, that gets the upper hand in Hume's thought. Hume's conviction in this regard deprived him from making sense of the world, including his own self, as a unity or as a whole.

However, and for the most part, empiricism has been understood as the bringing together, or the integration, of diverse experiences into a coherent whole. It is also the integration between thought and life without having to make use of *a priori* (what comes before the experience) and unrealistic metaphysical claims. In addition, and contrary to Hume's opinion, we do not experience things in an atomic way: we perceive things within a frame of reference. We bundle even seemingly disconnected perceptions and do the same with motives.[6]

Midgley insists that empiricism is not the opposite of the diversity of life or its complexity. On the contrary, it

---

[3] See Mary Midgley, *Beast and Man: The Roots of Human Nature* (revised ed. with new introduction; London: Routledge, 2002), 20–21.

[4] Midgley, *Wisdom, Information and Wonder*, 201.

[5] See, Anthony Flew, *A Dictionary of Philosophy* (2nd rev. ed.; New York: St. Martin's, 1984), 74.

[6] Midgley, *Beast and Man*, 265.

acknowledges the wide range of what constitutes the experience of the world, which is also to say that the world is perceived as varied and complex. There is no assumption that either diversity or complexity is necessarily or intrinsically wrong, or that knowledge based upon building recognition of such variety is necessarily confusing in principle. For Midgley, one good example of the empiricist mode at work is William James (1842–1910). According to Midgley, empiricism for James means to consider the wide range of experience "seriously and open-minded," and to ponder the best way to describe it.[7] It also means not to define the experience in advance in ways that rule out alternative descriptions.

Midgley finds James' empiricism very fruitful in the matter of dealing with religion. In James' seminal book, *The Varieties of Religious Experience*, besides attending realistically to those varieties, he also recognized how religion itself is an important part of the human experience and sense of the world.[8] Moreover, and in line with James, Midgley argues, "the intellectual attitude necessary for science, if given its full scope and not reduced artificially to a mere mindless tic for collecting, is continuous with a typically religious view of the physical world. This is one of the

---

[7] See Mary Midgley's discussion of James in *Evolution as a Religion: Strange Hopes and Stranger Fears* (revised ed. with a new introduction; London: Routledge, 2002), 130.

[8] Midgley, *Wisdom, Information and Wonder*, 201. In response to a question about how she thinks about religion, since religious concerns seem to frequently occupy Midgley's thinking, she wrote that she considers "how many elements there are in the various religions and what varying work they do in people's life. I don't think it's a single clear category at all. My Bible here (as you may have noticed) is William James's *Varieties of Religious Experience*." This quote comes from an electronic mail from Midgley to myself on August 17, 2005.

varieties of religious experience."[9] Here, by the way, Midgley sees another reason to avoid the misleading assumption that science and religion stand necessarily in opposition to each other. Religion is no less empirical than science when it considers the human experience of the world in both its outer and inner dimensions.

Getting Help from Darwin

Midgley believes that a more striking example of such a respect for the range of experience in the world is Darwin himself. Darwin's approach to the study of nature was successful because of his openness to the richness and variety that he encountered everywhere he looked. "Darwin himself was filled with a deep reverence for the world he studied, with the kind of awe and wonder that earlier philosophers such as Aristotle and Spinoza had described in religious terms as a response to its divinity."[10] Yet, Darwin did not rush into theorizing about what he observed; he did not allow theory to dictate firsthand his record and subsequent interpretation of the facts.[11] Theory came later. Darwin built literally from the ground up. He became basically a bottom-up thinker. Midgley points out that through Darwin's accounts of his voyage aboard the *Beagle*, we detect someone who was absorbed in wonder and awe of nature and its many life forms.

Nevertheless, Darwin's concentration on details and further reflection on these natural phenomena gave him eventually a sense for the relationships among very different creatures. His careful observations, accompanied by deep thinking on shared characteristics among many living

---

[9] Midgley, *Evolution as a Religion*, 131.

[10] Mary Midgley, "Darwinism and Ethics," *Medicine and Moral Reasoning* (ed. K. W. M. Fulford et al.; Cambridge: Cambridge UP, 1994), 14.

[11] Midgley, "Wisdom, Information and Wonder," 201.

things, gave him novel insights about possible patterns in nature, patterns that his contemporaries either failed to see or were unable to comprehend, or maybe were even unwilling to accept because of their possible implications for natural philosophy.

However, Darwin was not alien to or totally untouched by developments and discussions in the sciences of the day. Even Darwin held, as he himself acknowledged later on in life, some common assumptions about scientific method and doctrinal convictions. Thus, Midgley understands that parsimony in science, meaning reverting to the simplest possible explanation or the smallest number of principles, played a role for Darwin too.

> Though Darwin was so deeply concerned not to underestimate the existing richness and variety of life-forms, he still looked for a way of accounting for them that would be as clear and parsimonious as possible, that would invoke no mysterious underlying forces or tendencies. This quest for parsimony led him, as it had led other empiricists, towards an atomistic view, an account in which innumerable separate entities act each according to its own principles, and any appearance of common action or co-operation across the whole is delusive. The idea that what happens is in some sense 'pure chance' — is essentially contingent — does not appear to such thinkers as a hypothesis to be established, and often not as an obscure idea needing to be analysed, but as an unassailable presupposition guiding the whole inquiry.[12]

However, Darwin was not adamant in saying that this was the only possible conclusion that could be reached. He still allowed for the possibility that other kinds of factors were at play in the evolutionary process.

According to Midgley, the refusal to acknowledge any kind of deep connections or patterns among living organ-

---

[12] Midgley, 202.

isms, without having to revert to reductive explanations, has been the problem with many versions of evolutionary thinking to this day. The will to believe in an atomized world, where every living thing is a separate unit, independent from one another and with no reason for being except for its own individual sake, is still very much part of science as we know it. There is a conceptual failure to see how things can combine with each other for specific goals. For Midgley, therefore, "if... 'empiricism' is taken to mean a dogmatic belief that 'all beings in the universe, considered in themselves, appear entirely loose and independent of each other,' it is as remote from modern science as it has always been from everyday experience."[13]

This reluctance to even consider the possibility of organisms working together in concerted actions seems to be moved by fear of finding instances of purpose or aim in nature and not knowing what to do with it. Such reluctance actually flies in the face of the evidence. Nature shows many instances of cooperation among living things. This avoidance of "problematic facts" or "counter evidence" tends to violate the integrity behind the method of reasoning by inference to the best possible explanation after enough evidence has accumulated. In Midgley's view, the denial of all possibilities of purpose or direction in nature is not due to a strict application of empirical principles, but rather to an *a priori* refusal to even entertain such possibilities. It is the outcome of atomistic thinking, where "physicalist reduction" meets its "psychological" counterpart, and then tries to pass as science.[14] These are both forms of reductive thinking in science.

Moreover, the idea that evolution is best understood as an expression of the fierce competition of every living organism against the others, where each organism aims

---

[13] Midgley, 203.

[14] Midgley, *The Myths We Live By*, 43–44.

almost exclusively at its own advantage, is basically misleading. In reality, there are as many instances or more of altruism and cooperation in nature than of selfish or egotistic competition. The capacity especially among social animals to show an "active concern for others has—in several groups, separately—somehow become possible."[15] It is revelatory that a "problem of altruism" has occupied evolutionary biologists and especially geneticists for quite a while. The fact that such a "problem" is believed to exist, speaks volumes about the prior, non-empirical assumptions being made.[16] Since such an atomic view of things actually explains so little, especially in biology, it probably means that such a view owes more to social ideology, to one of the myths of the modern era, than to empirical science itself.

Questions About Method

Midgley sees much goodness, and not just usefulness, in building knowledge by moving from simple statements about the world to complex ones, not the other way around. She argues that since the Enlightenment, we have become used to an approach that begins with simple and sweeping concepts that are then impose upon the human experience of the world. This is in essence a deductive reasoning, which has become the very definition of the scientific method since the eighteenth century. The aim of utilizing such reasoning may sound trivial at first: organizing an otherwise confusing array of information from the world. But in reality, the tendency to overlook the complexity of natural systems, in favor of a simplified account of its working, has become a way of dispensing from the start with what is deemed not good: non-scientific or non-rational. "This melodramatic

---

[15] Midgley, *Beast and Man*, 127.
[16] Midgley, *Evolution as a Religion*, 143.

simplicity has been one of the chief theoretical attractions [of the Enlightenment]. It is also its chronic weakness."[17]

Part of the problem is that what seems to work in the physical sciences is then applied as the proper method of analysis in everything else. "Throughout the social sciences and often in the humanities, distorted ideas of what it means to be scientific and objective still rule much research. [However] even within science itself this approach is beginning to make trouble."[18] Unfortunately, and in spite of Darwin, this simplistic and overreaching approach still seems to linger. Biologists are still looking for simple mechanisms in allegedly truly scientific mode in order to explain the development of organic forms at the smallest level possible. It is a view from physics that seems to dominate here. Fortunately, a number of biologists are breaking with such a reductive approach and looking more to reconstruct "large-scale matters such as the behaviour of organisms in their environment."[19] The neglect of looking at larger units is no longer warranted. There is an effort to bring biology back into the science of evolution.

Despite appearances to the contrary, Midgley insists that Darwin's evolutionary idea was not a universalizing, overreaching principle. His was not an attempt to explain as the saying goes, everything from "gas to genius." Darwin actually rejected this kind of ideological imposition, which he thought was a sure sign of scientific overconfidence.[20]

For Midgley, the point of departure for any scientific method worth bothering with has to be the experience of life's complexity as it is manifested to us from the ground up. It is a movement from the simplest forms of organic life

---

[17] Mary Midgley, "Madness in Method," *Perspectives* [On line] 3/1 (1998): 1; the article may be fond online at www.mentalhelp.net.

[18] Midgley, 2.

[19] Midgley, 3.

[20] Mary Midgley, *The Owl of Minerva* (London: Routledge, 2005), 41.

and their interrelations to the more complex ones. Recognition of life's intrinsic diversity is not an obstacle to knowledge, it is constitutive of knowledge. Knowledge is ever a building process, something that is done through multiple projects and views. The need for explanatory principles, for their help in organizing the diversity of experiences and establishing law-like regularity, will inevitably come. Our intelligence has developed in part precisely by doing that, by making sense of that richness. What Midgley wants to avoid is the imposition of one way of thinking, of the one principle, the "one size fits all" approach.

> Because this complexity has now been recognized, we now know that it is vanishingly unlikely that a single way of thinking will ever explain the world's workings so thoroughly that we can dismiss all other ways as wrong, or as mere steps towards it. No one pattern of thought—not even in physics—is so "fundamental" that all others will eventually be reduced to it. Instead, for most important questions in human life, a number of different conceptual tool-boxes always have to be used together. And unfortunately, there is no single law showing us how we have to combine them. We simply have to keep on doing this carefully as the necessities of each case dictate until we reach a result that appears satisfactory.[21]

What is needed, among other things, is a way of looking at a thing "from below," from the innate pluralism of the world, and following the evidence to where it leads.

---

[21] Mary Midgley, "Concluding Reflections: Dover Beach Revisited," in the *Oxford Handbook of Religion and Science* (New York: Oxford UP, 2006); the quote is from page 14 of her original manuscript titled "Dover Beach: Understanding the Pains of Bereavement". By this means I want to gratefully acknowledge Dr. Midgley's kindness in sharing a copy of her unpublished essay for the purpose of this research.

Midgley considers herself to be in the Aristotelian-Darwinian tradition[22] of attending to the biological study of nature, including human nature, as indispensable in the way to knowledge of the world.[23] In the case of Aristotle, the philosopher still found a way to subordinate knowledge of the things below (the sub-lunar realm) to the knowledge of things above (the supra-lunar realm). Perfection was up there, while corruption governed down here. However, in the case of Darwin, there is no such hierarchy, except to say that knowledge of who we are and what is our home always starts (and for some exclusively remains) here.

Analogy with Theological Method

In Christian theology, a most contested issue is the one related to theological method. Major theological schools or movements throughout the 20th century have invariably

---

[22] See Mary Midgley, *The Myths We Live By* (London: Routledge, 2003), 128.

[23] On this point, I also rely on a response to a question that I posed to Midgley about points of agreement between her philosophy and that of Spinoza. For example, when it comes to such commitments as the role of emotions in thinking, the body-mind unity, the non-personal God, the practice of philosophizing as the pursuit of wisdom or as a spiritual exercise, the public and political role of philosophy, a critical engagement with science, and the role of religion, I told her that I find all these parallels between the two of them. Here is her answer: "Yes, indeed, I think that Spinoza and I do agree on all the points that you mention—which are all surely very important. But I haven't actually studied his works at all thoroughly... I suppose one could say that what he and I are both doing is agreeing with Aristotle (whom I have studied much more fully) on a lot of these matters...." As for Darwin, he is very much present and quoted through many of Midgley's writings, as may become clear through my presentation of her insights.

dealt with this challenge.[24] There is a sense in which the way of proceeding in theological exposition—what is considered first, or fundamental, or maybe even instrumental in the task of doing theology—determines much of its results, or where a theology eventually ends up. We could say that more often than not the way of "departure"[25] determines much of the way of "arrival."

Questions of theological method are often seen as 'prolegomena' (introductory remarks) to the theological task properly speaking. This is traditionally the way of letting the theologian's assumptions and commitments be known in advance, including some disclosure of where she or he gets his exegetical tools and method. A good example of how the way of proceeding affects the way of doing, or the way in which we look at some subject matter and, therefore, the way conclusions are reached, is the theological discipline traditionally called Christology. The latter has been a battleground and testing ground for diverse theological approaches as well as for questions in religious epistemol-

---

[24] To name but a few notable volumes, we have the German theologian Paul Tillich (1886–1965), who did much to explain and expand his "method of correlation" between theology and culture; see, for example, Tillich's *Dynamics of Faith* (New York: Harper & Row, 1958). In the Swedish school of interpretation, the most influential has been Anders Nygren (1890–1978), *Meaning and Method: Prolegomena to a Scientific Philosophy of Religion and a Scientific Theology* (trans. Philip S. Watson; Philadelphia: Fortress, 1972). And, in the American Catholic context, see Bernard Lonergan (1904–1984), *Method in Theology* (New York: Herder and Herder, 1972).

[25] Here comes to mind the Catholic Liberation theologian Jon Sobrino, especially his *Jesus the Liberator: A Historical-Theological Reading of Jesus of Nazareth* (trans. Paul Burns and Francis McDonagh; Maryknoll, NY: Orbis, 1993), especially 36–63.

ogy.[26] I for one have benefited much from this discussion in the past. Now it has also provided me with a way of looking at what Midgley is doing with questions of epistemology in evolutionary thinking. Because of its importance for my own way of naming what Midgley does, I need to add a few more comments on theological method.

It was particularly the late Roman Catholic theologian Karl Rahner (1904-1984) who identified what he saw as the two basic ways of constructing statements in Christology — that is, doctrinal statements about the person and work of Jesus Christ, and the present place and role of such doctrinal constructions in the thought of the Christian Church. The two basic ways of proceeding encompass two basic types of Christology: one builds knowledge up "from below," and the other, with a different point of departure, deducts knowledge "from above."[27] The former is more properly called a Christology of "salvation history," and the latter a "metaphysical" one.

Doing Christology "from below" means then to gather information, in this case the evidence from Scripture, about the so-called "Jesus of history," as portrayed in the written (canonical) Gospels. It begins with the testimony of the experience of Jesus as a human being — both his own and the disciples', but also ours, as we share with them a common humanity. Jesus is one of us. The bottom line is the existential element, the reference to an ordinary human experience, avoiding any temptation to add anything supra human or supra historical to the story of Jesus. Only that

---

[26] For more on this point, a basic reference volume is that of Danish theologian Jens Glebe-Möller, *Jesus and Theology: Critique of a Tradition* (trans. Thor Hall; Minneapolis: Fortress, 1989).

[27] See, Karl Rahner, "The Two basic Types of Christology," in *Theological Investigations*, volume XIII (trans. David Bourke; New York: Seabury, 1975), 213-223. Rahner was very influential predominantly in many Catholic theological circles but also in Protestant ones.

which is common to all is allowed to enter our thinking and consideration about the historical Jesus, a man from Nazareth.

The only assumption—safely made—is his common humanity, with its possibilities as well as its limitations. Beyond Christology, any theological reflection should likewise begin from what we know, or are able to know and believe comfortably, that is, experiential and sensible, or "from below," so to speak. The point of departure is what is common and familiar, for the most part, to us.

Now, doing Christology "from above" is to proceed from a conception of what could be the case, statements based on a form of philosophical realism, where the reality of a thing, its core nature or explanation, lies somewhat or at least partly beyond itself. It is a metaphysical understanding, presented as being more reliable than ordinary experience. In the case of Christology, when it is done "from above," it begins with convictions about the nature of the divine and of divine revelation. It speaks of the will and purposes of God. From such statements, conclusions are drawn about the ideal human experience, a sense of what ought to be the case, and not just what is the case. This approach is basically "deductive" in nature, since it draws conclusions from so-called axiomatic beliefs.

The latter does not imply that the former method, that of proceeding "from below," is necessarily inductive. It is a cumulative process, one that works by inference to the best possible explanation of the available evidence. It so happens that Darwin's method, as I (and others) understand it, is for the most part of this type.

Method Matters in Modern Thought

During the Scientific Revolution, the idea that what makes science rigorous and reliable was its method grew rapidly. Concerns with a proper method and secure way to attain true knowledge, although somewhat present already in

Greek science, actually reached a new paradigm in the early seventeenth century. The French philosopher and mathematician René Descartes is often cited as a major contributor to this process. In his influential work *Discourse on Method*, Descartes (1596–1650) details a program that includes the axiomatic use of first principles together with a detailed exposition of the deductive method of establishing true and certain statements of facts, of the kind that physical science has thrived on since. It is no accident that concerns about the reliability of knowledge and of certitude in the beginning of the Modern Era began with the work of a mathematically minded philosopher like Descartes.

> In at least four areas, in particular, there is ample evidence that [Descartes] was convinced that... his method deserved to supplant entirely what had gone before. First, he aimed to propound a unified scientific understanding of the universe, in contrast to the compartmentalized and piecemeal approach of the scholastics. Second, this science was to be based on mathematical principles, in contrast to the qualitative explanatory apparatus of his predecessors. Third... he wanted to develop a mechanistic model of explanation, avoiding whenever possible any reference to final causes and purposes... And fourth, the new comprehensive system of scientific explanation was to embrace, for the first time, the realm of human existence, including physiology, medicine, and a very large part of what we now know as psychology.[28]

Besides questions and debates about Descartes' own real or only perceived original contributions to the epistemological and metaphysical foundations of the new science, there is an attempt to establish systematic doubt as the proper

---

[28] See John Cottingham, "A New Start? Cartesian Metaphysics and the emergence of Modern Philosophy," *The Rise of Modern Philosophy: The Tension between the New and Traditional Philosophies from Machiavelli to Leibniz* (ed. Tom Sorell; Oxford: Clarendon, 1995), 148.

method of science.[29] Moreover, Descartes was one of the first to eliminate from science concerns about "efficient" and "final" causes, which were inherited from Aristotelian philosophy and that were still taught way into the seventeenth century academic curriculum.

For all of Descartes' successes, it is not really until the work of British physicist and mathematician Isaac Newton (1642–1727) that we get a proper account of what a method for the natural sciences looks like. Newton began the second volume of his seminal work, the *Principia Mathematica*, with a list of principles for true reasoning in what was then called "natural philosophy."[30] He presented his principles in three basic, successive rules, structured in the form of axioms.[31] In them, Newton presented his own version of what has been known to this day as "Ockam's Razor," which states that the simplest possible and successful explanation should hold and be enough. For Newton, what mattered was not to admit more causes for natural phenomena than necessary. Moreover, the same causes should always be used, as long as they serve to explain equal or similar effects or phenomena. Finally, all qualities of bodies, attested by experimentation, should be considered universal qualities of such bodies. Newton's impact on the scientific method has been tremendous.

Newton's view of the universe was of a universe in motion. His understanding of physical reality was basically atomistic, but not in the old sense. Newton's atomism did not entail a universe determined by chance and accident. His universe was a balance of forces and objects in motion

---

[29] Cottingham, 149.

[30] Isaac Newton, *Mathematical Principles of Natural Philosophy*, 2 vols. (original ed., 1689; trans. Andrew Motte, 1729; rev. by Florian Cajori; 6th printing; Berkeley, Cal.: University of California, 1966).

[31] See Newton, 398.

through space and time that is ordinarily regulated by natural laws which are mathematically defined, not merely by us, but established in this way by God himself. It is a world of discovery. And these natural laws, the mathematical principles of nature, govern equally the vast universe as well as the small realm of atomic particles.

Since Newton's time, mathematically oriented science has been used successfully in order to explain the physics of the world. All physical reality has been studied by dividing it into its smallest parts or constituents. Atomistic thinking has ruled in the physical sciences ever since.

> Physics wins here because it stands nearest to the end of the quest that dominated science from the time of Galileo until quite lately: the atomistic project of explaining the behavior of matter completely by analyzing it into solid ultimate particles moved by definitive forces, "ultimate building-blocks," as people still revealingly say.[32]

The proven success of this way of natural analysis has been also tried repeatedly in other disciplines. What some have termed (ironically) "physics envy"[33] has influenced the way every scientific study is conceived.

> What finally broke up this hospitable, inclusive idea of science was the steadily growing prestige of the physical sciences themselves. That prestige led a growing number of prophets to jump on the bandwagon by claiming that their ideologies were scientific—not just in the old sense of being methodical but in the new one of being founded in some way on physical science.[34]

---

[32] Mary Midgley, *The Myths We Live By* (London: Routledge, 2003), 33.

[33] Mary Midgley, *Science and Poetry* (London: Routledge, 2002), 132; I don't think that the phrase "physics-envy" originated with Midgley, but she has certainly used it a number of times in her writings.

[34] Midgley, 149.

Everything from organic life to the human mind and psychology has been and is still being dutifully analyzed by dividing it first into individual units or components. The idea is to be able to explain what anything is by study of its smallest parts. By deduction from what seems to be most basic and elementary, conclusions are drawn about what the whole is, without really having to look at the whole—as whole—in the first place. This is classic reductivism, which is fundamentally mathematical in character. However, what has worked so well for a mathematical physics, and especially for particle physics, does not necessarily work in other areas or with other objects of inquiry.

From what I can tell, biological science has resisted for the most part this kind of procedure. First, biology deals with life as a complex phenomenon. It means to acknowledge the immense variety of living forms, each with its own peculiarities. Second, life is an emergent process, not something of which we could exactly say when and where or under what conditions it actually began. Certitude is lacking here. Mathematical analysis has proven insufficient to account for the complexity of life forms in their unique environments and in an evolutionary process that has lasted for millions and billions of years, beyond our capacity to comprehend, let alone reproduce in all its details and steps. Third, and finally, despite the great number of studies and amount of information collected on the evolution of life on earth, it is only thus far a minuscule part of the potential of such information that has been gathered. It is normally held by biologists that a great majority of all species of plants, animals and bacteria that have ever existed have not survived. What is part of our present knowledge might not be more than, in some estimates, ten percent of all life forms that have existed in the earth's history. Other estimates make the number of extinct life forms an astonishing 99.9

percent of all organisms that have ever lived.[35] Indeed, it would seem a lot less than 10 percent are known, since perhaps most species of living creatures are not even known either.

How then can the science of living things (biology) establish facts or builds up knowledge of the world? It does it by the accumulation of whatever evidence is available (lots of it, by human standards anyway), and by proposing solutions to the many enigmas through induction, by use of the best possible explanation for the available evidence. It normally reaches this stage by what analysts call the "tipping point." The latter is a way of saying that the accumulation of data will inevitably reach that point at which an explanation will become clear; it will offer itself to the mind, so to speak.

For the most part, the science of evolution seeks explanation by induction. This is not to say that deduction plays no role. As a matter of fact, Darwin himself thought at times that "natural selection" was such a principle, like a first principle, from which a number of deductive conclusions follow. However, at other times, Darwin insisted on the metaphorical character of "natural selection." What began as the search for a gravitation-like law,[36] ended in a proposal as to the theory that can best explain the evidence until this point in time.

Despite Darwin's confidence that natural selection, and its companion mechanisms of variation and adaptation, were solidly established and confirmed beyond doubt by the evidence collected during his time, he still had to admit that not all the evidence was yet available. There were gaps in the understanding of the evolutionary process, as well as

---

[35] See, for example, Ernst Mayr, *What Evolution Is* (New York: Basic Books, 2001), 140.

[36] In the concluding pages of the *Origin of Species*, Darwin was still referring to this hope of having found laws such as laws of nature; see 648.

gaps in the geological record. Besides, there was also the possibility that other parallel principles, though probably not as significant, might be at work but yet unknown to him and his contemporaries.[37] Midgley understands the difference between inductive reasoning in evolution and deductive reasoning in physics. Multiplicity of methods and approaches is not a problem. It stands as a testimony to pluralism in science, in method as well as in content and scope. The real problem comes when the methods of one science, physics, are thought to be universal. When mathematical thinking is indiscriminately applied to other disciplines or fields of inquiry, it invites disaster. It is conceived in reductive terms only. By aiming at clarification, by limiting reasoning to distinct statements of facts, it actually reduces significantly what counts as knowledge. According to Midgley, even the English empirical tradition in the hands of seventeenth and eighteenth century thinkers suffered this fate. For instance,

> Hume's rules [of experimental reasoning] allow meaning only to mathematics and to reports of sense experience; everything else is dismissed as nonsense. He does not tell us what he thought ought to be done with... the rest of empiricist philosophy. This was an early form of the reductive distortion that caused so much trouble later, the hasty adoption of a bizarrely restrictive view of meaning to shore up a shaky metaphysical position, without proper attention to the problem of what "meaning" means.[38]

As Midgley implies, this kind of reductive distortion happens also in biology when atomistic thinking is applied, or

---

[37] See Darwin, *Origin of Species*, 213, 406, and 616. In those pages we find references to the problems of "extinction" and substitution of species, the "geographical distribution" of species, and questions about the full impact of "climatic—or environmental—changes."

[38] Midgley, *Wisdom, Information and Wonder*, 199.

rather misapplied there. It creates such aberrant results as selfish gene imagery, modular brain cognition, and also the idea of "memes" as evolving cultural units. The same kind of reductive approach has also been tried with religion, as in the language used to describe transmission of religious ideas as "viruses of the mind."[39] Atomistic thinking and its method can also be seen in the case of psychological individualism, where the self is said to be but an illusion, a conglomerate of unrelated conscious activities, or of diverse and unconnected psychic experiences.

Midgley Warns About Reductivism

One of Midgley's deep philosophical concerns has to do with the uses and abuses of what is called the "reductive approach" in science. Midgley understands the argument that consistently applying a reductive method to all its problems is what science basically does, and that that is what makes it unique among other forms of human inquiry and knowledge, but she also sees what reductivism does to science, namely, that it blinds science to possible solutions coming out of other disciplines. Midgley insists that a reductive approach is not merely a practice; it is rather an attitude.[40] Wholes are reduced to parts, different conceptual schemes are all reduced to one unifying principle or forced into one simple explanation overall. Although these two

---

[39] This expression is originally attributed to the Oxford biologist Richard Dawkins, "Viruses of the Mind," in *Dennett and his Critics: Demystifying Mind* (ed. Bo Dahlbom; reprint ed.; Oxford: Blackwell, 1994), 13–27. Since then it has been much used and debated. One good analysis of the possibilities and excesses of such an approach, while critical to its application on theological ideas, is John Bowker, *Is God a Virus? Genes, Culture and Religion* (London: SPCK, 1995), especially 37–46.

[40] Mary Midgley, "Reductive Megalomania," in *Nature's Imagination: The Frontiers of Scientific Vision* (ed. John Cornwell; Oxford: Oxford UP, 1995), 133.

examples may look essentially different at first, they also seem to have something in common: they both reveal intentions. Probably the best-known example is that of "Ockam's Razor," a principle of ontological economy, which counsels not to multiply entities or explanations beyond strict necessity.[41]

Formal reductions do not come in naturally. They are constructed as tools, as elaborate approaches. In other words, they are conceptualizations. And they are used as strategies within larger systems of thought.

According to Midgley, reductivism—the overuse and therefore abuse of a reductive approach—should not be confused with the legitimate use of parsimony.[42] Parsimony is about getting the most economical explanation possible that will still provide us with a correct answer to the question being considered. However, parsimony does not lead to reduction for reduction's sake. Even parsimony can be invoked for the wrong motives, like when it is used to avoid working through complex ideas or experiences. In any case, when reductivism becomes the sole rule and direction, what we get is a systematic exclusion of all other alternatives by principle. It is this latter case that Midgley calls the practice of "reductive megalomania."

Now, some kinds of reduction are appropriate. For example, in the relationship and possible overlap between two disciplines: Midgley mentions the case of chemistry and physics. One (physics) picks up the explanation of the "behavior" of matter where the other (chemistry) leaves it. In this case, physics seems to be more fundamental than chemistry: it seems to go deeper, at least in terms of the ultimate composition of matter. But chemistry has also benefited from the relationship. It is sometimes the case that one

---

[41] See basic definition in Antony Flew, *A Dictionary of Philosophy* (revised 2nd ed.; New York: St. Martin's, 1984), 253.

[42] Midgley, "Reductive Megalomania," 134.

discipline is seen through the lens of another; imposing the rules or methods of one science upon another may actually distort the understanding of the two.[43]

For example, physics has for the most part enjoyed such a priority of status, whereas other disciplines have been submitted to the methods of physics. This has happened to biology itself, with both chemistry and physics imposing on its methods and approaches, to the consternation of those biologists[44] who struggle to keep alive what they rightly conceive as the bigger picture of the complex development of life on earth. This is the case with evolutionary biology: it is the "bigger picture," the so-called "macro-evolutionary process," which serves it best. The difficulties for evolutionary biology start at the micro-evolutionary level, where questions brought in from chemistry (or biochemistry) and physics lead the way.

Is An Evolutionary Epistemology Possible?

Midgley's methodological proposal, which requires that we judge the basics of human nature and cognition from an earthly perspective, is her understanding of what an evolutionary epistemology can do for us. She herself has agreed

---

[43] Midgley, 135-136.

[44] In this context, Midgley mentions Francis Crick, *What Mad Pursuit* (London: Penguin, 1989), 139, where Crick complains that "physicists are all too apt to look for the wrong sorts of generalizations, to concoct theoretical models that are too neat, too powerful and too clean… To produce a really good biological theory one must try to see through the clutter produced by evolution to the basic mechanisms lying beneath them… What seems to physicists to be a hopelessly complicated process may have been what nature found simplest…." I would add the book by Steven Rose, *Lifelines: Life Beyond the Gene* (Oxford: Oxford UP, 1998), where keeping the broader picture in the complexity of relationships and the dynamics between genes and environment forbids any reductive understandings.

with this assessment. In her response to one of my questions, she stated that "in asking what we know, I start by asking about ourselves as knowers, and as beings who are shaped by a particular evolutionary history, rather than by trying immediately to understand the subject-matter that we know (or don't know) about." Moreover, she added: "I'm not sure that I have spelled out this point about epistemology myself in my books."[45]

With Midgley, there is no doubt in my mind that Darwinian evolutionary theory has some profound epistemological consequences. Due primarily to its understanding of nature and its mechanisms as foundational to the origin and development of all life, it allows us to consider seriously naturalistic explanations to very divergent phenomena. Cognition, being a human process, may be assumed to have developed evolutionarily, like anything else human.[46] Scholars working on diverse disciplines and of different persuasions, tend to use evolutionary language as an analogy when applied to many other fields outside of biology.

However, naturalistic explanations are not the only ones available. Moreover, they should not be necessarily understood as opposite to metaphysical ones. It is also true within this scenario, that the move from naturalistic to metaphysical conceptions is basically inevitable. We are to follow the lead, as it is so often stated, wherever the evidence may take us. If the possibility for metaphysical explanation is ruled out in advance, we will normally fall short of getting a complete picture.

---

[45] These quotes by Midgley come from our personal exchange and her response through electronic mail, more recently on 02-28-06. I have kept copies of all our correspondence for study and reference.

[46] See, for example, Ton Derken, "The Promise of Evolutionary Epistemology, Its Coherence and Its Relevance," *The Promise of Evolutionary Epistemology* (Tilburg: Tilburg University, 1998), 1.

In the case of Darwin, he seems to have had a rare combination of attention to details and the concern for the big questions. With him the inference from the evidence to the best possible explanation and, at the same time, working with predictions and testable theories, were almost of one piece. As American paleontologist Niles Eldredge explains, in the language of scientific methodology,

> Darwin, rather consciously, switched gears from a predominantly, more-or-less Baconian inductionism — which got him, as he himself saw it, to the conviction that life has evolved. But as he later said, this was not wholly satisfactory: he wanted to discover just how species change through time. Once grasped, natural selection became the tail wagging the dog in the sense that all the patterns Darwin originally saw were, in essence, reformulated as predicted outcomes of the selective process.[47]

Moreover, according to Eldredge, "it must be said that Darwin did not play exactly by the rules of the hypothetico-deductive method," basically because Darwin never rejected what he considered good hypotheses when these did not fit well with the data or his predictions. He was rather confident that further data might clarify the missing point and bring back explanations that had been put on hold. In brief, "Darwin's switch to a hypothetico-deductive stance once his inductionism led him to the truth of evolution (no mean feat!) seems admirable, and certainly a good

---

[47] Niles Eldredge, *Darwin: Discovering the Tree of Life* (New York: Norton, 2005), 94. This book is a companion volume to the catalog of the Darwin exhibition that celebrates Darwin's two hundredth birthday (he was born in 1809) and the one hundred fiftieth anniversary of the original publication of the *Origin of Species* (in 1859). The museum exhibition begun on November 2005 at the American Museum of Natural History in New York City, and eventually moved to four major world cities, ending in London in 2009 in time for the big celebrations.

rhetorical and intellectual stance to take when making his case for evolution through natural selection."[48]

As we can see from Eldredge's own explanation of what Darwin was doing, Darwin's approach suffers from a kind of circularity. For example, if we take natural selection as a metaphysical principle giving direction and as a guarantor of the organic evolutionary process as a whole, then what we see is the use of a metaphysics for the theory to support the very evidence that we already assume is providing the relevant proof, as building the case for evolution. Our overall assumption here would then be that the theory of evolution takes for granted the realism of the facts that it accumulates and helps to interpret. The theory of evolution adds ontological status to the world, the world as it is, while at the same time wanting to assert the primordial epistemological nature of the theory.

By implication, after Darwin became convinced of evolution by accumulation of evidence (induction), he then retroactively applied evolution deductively to his research examples. Thus, in a sense, Darwin was inevitably finding what he wanted. It seems though that for Eldredge, the main point is to prove Darwin's eclecticism in regards to method, or his plurality of approaches.

In its further history, Darwinian evolution has eventually broken with a more traditional understanding of metaphysics.[49] One example of that move is the central position that the individual organism has come to occupy in the competition for survival. For Darwin, species are primary competi-

---

[48] Eldredge, 95.

[49] See, for example, the arguments in this regard by Michael T. Ghiselin, "The New Evolutionary Ontology and Its Implications for Epistemology" in *Darwinism and Philosophy* (Notre Dame, Ind.: University of Notre Dame, 2005), 245. By a traditional metaphysics is meant the attempt to characterize reality as a whole instead of particular parts or aspects of it; the latter is commonly held to be the approach of the natural sciences.

tors. However, species are also classes of populations, while populations are made up of individuals. But can species be treated as empirically real, not just conceptually so? In principle, a species is a class, not an individual. For a realist, like Darwin himself, however, both are real. Still classes seem abstract while individuals are concrete. The contemporary emphasis on selfish gene behavior, for example, has been born out of this emphasis on individual roles over populations.

Ever since Darwin, on the other hand, evolutionary biology has looked for laws of nature in classes of populations. The reason for this is that there are no laws of nature exclusively working for individuals strictly speaking.[50] Nature does not create specificity in order to serve individual needs in exclusivity. Therefore, as a theory of organic processes in the world, Darwinian evolutionary theory, although mindful of these difficulties, cannot avoid working at multiple levels: individuals, populations, species, and even life as a whole. Why? This is because Darwinian evolution is at heart a historical perspective. It has helped to make clear that biological science is fundamentally historical knowledge, in which explanation by historical narrative is the ultimate goal and criterion for evaluating its truth-claims and knowledge. In solving biological problems, the appropriate questions and answers are likewise historical.[51]

Being basically historical in nature, the evolutionary narrative explains that things have happened in a certain way and dares to hypothesize about the possible causes of why they happened that way. In fact, all claims as to the possible causes of biological traits in organisms and their

---

[50] Ghiselin, 250.
[51] Ghiselin, 251.

heredity entail an explanation by historical narrative.[52] However, there is always doubt about whether Darwinian theory can appropriately allow for teleological explanations. Being fundamentally a naturalistic perspective, it does not fit well with teleological explanations, which are said to be basically metaphysical, dependent on ontological assumptions about progress and direction toward a predetermined goal. As Michael Guiselin (b. 1939) suggests, Darwinian theory cannot deal with occult causes (like foresight, for example) beyond accounting for what the evidence shows about the possible history of organic life.

As far as we can tell, all human cognition has developed from less toward more elaborated forms through the evolution of organic life. Scientific cognition, as a case in point, cannot be presumed to be working in a way necessarily different or independent from other human processes. Science itself, as scientist and humanist John Ziman (1925–2005) has forcefully explained, is social.[53] It is a product as well as a shaper of our present experience of the world. The capacity

---

[52] Ghiselin, 252. Guiselin mentions a counter example to this position, namely, the arguments of philosopher Daniel Dennett, who treats the evolutionary mechanism of adaptation from an ahistorical point of view, or as a general explanation that could be inferred, and therefore applied, from general natural laws or principles. However, neither Guiselin nor Dennett makes clear whether a non-historical argument from necessity is ever a conclusive one for evolution. See especially Dennett's book, *Darwin's Dangerous Idea*, 238 and 250.

[53] See John Ziman, *Real Science: What It Is, and What It Means* (Cambridge: Cambridge UP, 2000), x. Ziman began his professional career as a scientist (physics), but after a couple of publications in the field, turned his attention and rigorous analysis to philosophical questions regarding science. In his published works, he explored the social and psychological dimensions of scientific practice. He was a pioneer in what is now known as the sociology of knowledge, among other areas of research. My inclusion of his work has been done following Mary Midgley's own advice in her response on February 28, 2006

for human knowing, in all its forms, is the product of both biological and cultural (social) evolution.

Knowledge of the physical world — its geometry, relations between objects, reliability of most perceptions — all of these are encoded in our genetic make-up and not just recorded in our memory. This knowledge has gone through the same process of selection, variation and adaptation as anything else organic. We can say that our cognitive capacities have developed in response to the world and our experience of it. In this sense, current cognitive faculties can be said to be "innate."[54] In an evolutionary perspective, this is how we have reproduced the world to ourselves, as needed for survival. What has worked well for us thus has also become "hard-wired" in our cognitive systems. For example, we are said to be adapted to expect the regularities of the world.[55]

The above assumptions are basic to an evolutionary epistemology, which asserts that knowing the world from where we have our origin and where we now live is intrinsic, not peripheral, to human life. The human knowing process works with the world and all its entities.[56] Everything in the world plays a role in forming cognitive functions and, therefore, contributes to knowledge.

One caveat to these functions and their formulations is that they are neither perfect nor optimal. What matters is that they have been sufficient for survival. The kinds of representations of the world that this complex of cognitive

---

       to one of my electronic letters. When it comes to the question of epistemology, and of evolutionary epistemology in particular, she says she is in basic agreement with Ziman's views.

[54] Ziman, 299. Midgley criticizes attempts at denying some inner instincts and faculties in human nature; she does it by opposing "blank paper" theories of human knowing because these tend to treat human beings as "totally plastic and structureless." See Midgley, *Beast and Man*, 18–19.

[55] Ziman, 300.

[56] Ziman, 300.

faculties and experiences produces are like maps for a territory, maps that represent different aspects of the world. Human beings owe much of their success as organisms to the further evolution of more complex cognitive capabilities, such as recognizing patterns, defining similarity classes, constructing "maps" and mental models, and transforming these socially, through communication, into intersubjective representations. In other words, various models of practical reasoning which are fundamental to science emerged originally as cerebral tools for coping with the hominid life-world.[57]

Therefore, "what evolutionary epistemology does demonstrate... is that this operation is self-consistent, socially coherent, and inclusive of both the natural and humanistic aspects of life."[58]

As we stated before, and in consonance with Midgley's thought, Darwin's theory of organic evolution has important philosophical consequences, of which some are epistemological. As far as we can tell, all our cognitive functions are evolved capabilities. The brain, which is the base of all thinking and cognition, originated through organic evolution. Our cognitive functions seem well developed for our environment, but they are not necessarily optimal; they are sufficient but not ideal.[59] Cognitive development is not about perfection but effectiveness. The bottom line is, evo-

---

[57] Ziman, 300; the author introduces in his writing the concept of "life-world" to speak of the world of everyday knowledge, which is actually rich in its multiplicity, diversity and heterogeneity, the environment in which organisms, to use evolutionary language, survive and reproduce; see also 152, 296, and 299.

[58] Ziman, 301.

[59] On this point, see Gerhard Vollmer, "How Is It That We Can Know This World? New Arguments in Evolutionary Epistemology," *Darwinism and Philosophy* (Notre Dame, Ind.: University of Notre Dame, 2005), 260.

lutionary epistemology tries to explain not only what works in human cognition but also what does not work. Epistemology deals with both the extent and limitations of our knowledge of the world.

However, evolutionary epistemology inevitably makes some basic assumptions about that same world that it is trying to explain. For example, that there is a real world independent in its existence from our consciousness of it—this assumption is what is normally termed "ontological realism." This kind of realism assumes that this world is, at least partially, knowable and therefore intelligible. The world remains an open object of our thinking and analytical faculties, or what is termed "epistemological realism." Epistemological realism then understands that our knowledge of the world remains simultaneously hypothetical and incomplete, in a sense preliminary, resulting in what the philosopher of science Gerhard Vollmer calls "epistemological fallibility."[60]

Realism is a fundamental premise of natural science. Although the success of science is not irrefutable proof of the truth of realism, it is nevertheless an argument for it.[61] In any case, by the assumption of realism we are able to explain more by science than through competing views of reality, whether forms of idealism, spiritualism, or social constructionism. In empirical science, the explanatory power of theories matters much. This is especially the case with those approaches based on a realistic conception of the world, something that has proven to be quite successful by these standards in the natural sciences.[62]

How do scientific assumptions about reality and the eventual successes of the natural sciences, play in Midgley's thought? For Midgley, the "real" cannot be strictly limited

---

[60] Vollmer, 262.

[61] Vollmer, 269.

[62] Vollmer, 269.

to physical objects or "atomic" (elementary) units in the world. Conceptual thinking, the inner or subjective experience, and moral values are all as real as any other object of conscious activity. As she likes to say, a toothache is as real as a tooth. An ontology restricted to physical and quantifiable objects makes things very difficult to understand. It makes the world not more but actually less intelligible. And it denies the factual nature of consciousness itself, as well as of ethical values and religious beliefs. It tends to make of the latter kinds of experiences "problems."[63]

Bias Against Earthly Things

Midgley has written much about the relationship between humanity and nature. In principle, a dichotomy between humans and the rest of nature is unsustainable. We are part of nature, neither above nor under it. However, ideas and beliefs about the opposition between earth and heaven have played an important role in separating ourselves from the rest of it.[64] A case in point is the conviction that our stay on earth is only temporary and therefore a prelude to heaven, heaven itself being our final destiny. Midgley believes that this idea has been at the root of some unhealthy notions in Christianity itself of what the earth is and what our place in it is too. Ironically, during the European Enlightenment, although secular thinkers abandoned the idea of heaven as destiny, the lack of clarity in thinking about the earth as our true home continues. A renewed sense of ourselves in rela-

---

[63] On this point, I rely partially on Midgley's arguments in her essay, "Dover Beach," quoted above in note 21 of this chapter.

[64] For this discussion, see Midgley et al., "Environment and Humanity: Friends or Foes?" sponsored by the St. Paul's Institute of St. Paul's Cathedral in London, on September 21, 2004, with the participation of Dr. Mary Midgley, Dr. Roberto Navarro, and the Archbishop of Canterbury, Dr. Rowan Williams. For a full transcript of the conversation, follow the link at <http://www.stpauls.co.uk>

tion to the whole of life on earth did not effectively sink into us. Thus, "people stopped thinking of themselves as souls that would go up to heaven. Instead they thought of themselves as minds which are so clever that they can control everything and are somehow independent of the higher sphere in which they live."[65]

According to Midgley, since the seventeenth century we have moved from the detailed study of nature as creation and as a manifestation of the work and goodness of God to the belief that nature stands better by itself, on its own terms, without need for reasons or beliefs considered external if not antithetical to nature. Nature alone is worthy of all attention. There is no doubt that our ideas of what nature is have changed.

Mindful of opinions contrary to her views, Midgley likes to point out that a tradition deeply rooted in Christianity has held a bias against things terrestrial. For long, everything that belonged down here, as opposed to up there in the heavens, has been looked upon with suspicion. Things terrestrial have been termed inferior, secondary, or even superfluous. The Earth itself has been seen as the realm of everything temporal—unlike the heavens where eternity belongs—and therefore as being limited, imperfect, and not really worth that much. The ground of the earth, its dirt, has been synonymous with unclean, corruptible, and messy stuff. Dirt has been an image for things putrescent,[66] and even for sin and sinfulness.

On the upper side then stand the heavens, the realm of light and purity. From biblical times and up to the present, the heavens have been considered as the sphere of things

---

[65] Midgley et al., 3.
[66] See, Mary Midgley, *The Myths We Live By* (London: Routledge, 2003), 122. Midgley consults the Oxford English Dictionary and finds as the meaning of "earthy", "heavy, gross, material, coarse, dull, unrefined... characteristic of earthly as opposed to heavenly existence..."

divine, the place of God's dwelling. The heavens announce both the beauty of creation and the wisdom of God. For Midgley, this dichotomy between heavens and earth, between upper and lower realms, has meant the ruin of the earth and of its "lower" creatures (lower than the human, that is).

Now, Midgley does acknowledge different strands in the Christian tradition regarding our relationship with nature and all other living creatures. One strand of tradition that has been alive in Christianity is the one that points to the need for the care of creation.[67] It speaks of our calling to be good stewards of the earth. But even this notion of a steward she finds somewhat patronizing.[68] And I may add, it has been mixed with notions of "dominion" over the earth, as in the end of the first account of creation in the Book of Genesis.[69]

Moreover, the Christian tradition has always spoken of the earth as creation, and as being therefore God's own. We can only receive it as gift, never as owners. This doctrine no doubt is out there to foment a notion of reverence toward the rest of creation. But Midgley wonders how truly influential in practice that teaching has been, since many Christians throughout the ages have rather spoken of earthly matters as corruptible, temporary and, therefore, basically unreliable. Earthly has stood for ungodly or unspiritual. It seems to her that the cultivation of a true natural spirituality — a concept that unfortunately Midgley

---

[67] Midgley et al., "Environment and Humanity," 6.

[68] Midgley et al., 15.

[69] After the creation of humankind, as man and woman, "God blessed them and said to them, 'Be fruitful and multiply, and fill the earth and *subdue it*; and *have dominion* over the fish of the sea and the birds of the air and over every living thing that moves upon the earth.'" (Gen 1:28); emphasis is mine.

does not explain — has been rather more absent than present in the history of Christian beliefs.[70]

Midgley is of the opinion that more often than not this negative view of things terrestrial has dominated Western religious thought. As with any belief or mode of thinking, there have been some notable exceptions — we have already mentioned St. Francis of Assisi before in this regard.[71] However, when bias against nature has mixed with fears of pagan religiosity, its negative effects have spilled over into social contexts as well. It has been the cause, or at least the excuse, for the persecution of witches and others who practice naturalistic forms of religious rites.

Then Darwin came and connected us all closely with other animals. By doing so, he also made of the earth our only true home. It is that unity with the rest of life, with all living things, that for Midgley is Darwin's most lasting contribution to our scientific as well as our religious thinking.[72] For Midgley, this point about Darwin's emphasis on our basic earthly nature and connection to and dependence on all other living creatures, is especially relevant today as we are living in a time when ideological misconceptions about the earth and its creatures have reached dangerous levels. The environmental crisis, notable in the increasing disappearance of green areas, global warming, the significant reduction of biodiversity and the fast extinction of countless species, has raised the specter of the impossibility of keeping life in this planet as we know it for much longer.

---

[70] Midgley seems to follow the conventional wisdom, at least about Christianity. At one point, she even suggests that this bias against things natural may come from Patristic thinking. Unless she is able to prove the point textually, I cannot follow her opinion in this regard.

[71] See the second chapter of this work.

[72] Midgley et al., "Environment and Humanity," 3.

Midgley's insistence that Darwin has helped to drive the idea home that we are of this earth, being her products, that we belong here, are at home, and that this is a proper way to tackle issues in evolution and religion. She insists that, "the crucial point is that life is not an accident or an alien invader but an aspect of the earth itself."[73] She has actually suggested that Gaia theory might be the best conceptual tool available in order to engage the difficult question arising from the relationship between religion and science, or science and spirituality.[74] Midgley claims that Gaia theory is not only capable of provoking a sense of awe and gratitude in us, but moreover it helps to convey "the sense of life as active and effective... made far stronger, not weaker, by our grasp of evolutionary theory."[75]

Midgley's philosophy of science and religion helps us understand that evolutionary theory and religious views do not have to engage in a "winner-takes-all" kind of competition. On the one hand, evolutionary theory can be defended as a reasonable, even provable, explanation of the development of organic life on earth. On the other hand, a religious view of reality can be safely defined as a belief in God or at least on an ultimate reality, which is taken as being foundational to all reality, in a metaphysical sense. Neither one is a substitute for the other, despite all modern attempts, and some notable examples, to make it so.

A Sense of Wonder

It is my conviction that we often fail to see the deeply rooted concerns that unite scientific and religious attitudes towards the world. As a general rule, we are more prone to

---

[73] See Mary Midgley, *Gaia: The Next Big Idea* (London: Demos, 2001), 40.

[74] Midgley, 21–23; see also, Midgley, ed., *Earthy Realism*, introduction, 3–9.

[75] Midgley, 24.

see or look for points of conflict between these two views than to find commonality. Our general tendency is to measure the success of religion by the methods normally attributable to science, leading to the result that we then find the former faulty or wanting. And if we do it the other way around, we may reach a similar result.

In Midgley's opinion, a better way of addressing this issue is to look at the similarity of motives behind both efforts (science and religion) of the human spirit, especially, the sense of wonder.[76] Wonder is surely a common experience in religion: a reverence for God, for life, for nature, or for mystery itself, whatever may be considered greater than ourselves. In the case of science, although it is commonly accepted that curiosity is a major motivation, a genuine sense of wonder is never far off. Curiosity is often driven by a sense of awe. It is not uncommon to hear from scientists themselves how they can feel moved to reverence—or at least to a deep respect—for the world by their own subject of study. This asseveration seems rather commonplace among physicists and cosmologists. Either because of the immensity of the macro-cosmos, or the complexity of and the perplexity provoked by the micro-cosmos, many scientists have testified to a sense of awe and beauty arising from the world and its explicit or subtle order.

One prime example is Darwin himself. What I have in mind is his most famous passage from the *Origins of Species*. In the conclusion to the work, he writes:

> It is interesting to contemplate a tangled bank, clothed with many plants of many kinds, with birds singing on the bushes, with various insects flitting around, and with worms crawling through the damp earth, and to reflect that

---

[76] See Mary Midgley, "The Need for Wonder," *God for the 21st Century* (Philadelphia: Templeton Foundation, 2000), 186. Although Dawkins has made a similar point, his view excludes religion and argues that science alone suffices because for him only science deals with true statements about the world.

these elaborately constructed forms, so different from each other, and dependent upon each other in so complex a manner, have all been produced by laws acting around us... Thus, from the war of nature, from famine and death, the most exalted object which we are capable of conceiving, namely, the production of the higher animals, directly follows. There is grandeur in this view of life, with its several powers, having been originally breathed by the Creator into a few forms or into one; and that, whilst this planet has gone cycling on according to the fixed law of gravity, from so simple a beginning endless forms most beautiful and most wonderful have been, and are being evolved.[77]

If this cannot be called a sense of wonder and the spiritual, then I do not know what to call it.

The sense of wonder has served as a point of departure, that is, the reason for doing science in the first place, and as a point of arrival as well, meaning that understanding itself increases the awe. These are the kinds of scientists who do not deny getting a sense of the spiritual from what they do. Their experience may be called "spiritual" as long as they convey some sense of belonging to a greater whole. It means that what they have discovered, unveiled or simply come to know is not the whole of the story; there is always more than "meets the eye." On the one hand, knowledge increases by addition of facts. On the other hand, new sorts of explanation surface now and then. But, through our taking up these new explanations, it is our vision of the universe that changes.[78] Our worldviews are challenged or corrected or just confirmed. Facts are not accumulated just for any reason: they are sought out to explain something, or to get to know better that which surrounds us.

---

[77] Charles Darwin, *The Origin of Species, by Means of Natural Selection, or the Preservation of Favored Races in the Struggle for Life* (original ed., 1859; reprint ed.; New York: The Modern Library, 1998), 648–649.

[78] Midgley, "The Need for Wonder," 188.

Today's scientists have caught the public's imagination by their descriptions of what is real beyond our wildest dreams. Midgley speaks of a mix of astonishment and awe.[79] Science's pronouncements appear to us and move us in the way religious sentiments do. Religion likewise is "inspirited by imagination," but also by openness to fresh ideas or forms of explanation.[80] The idea that reason and intelligence do not play a significant role in religion, or that imagination does not play a significant role in science is not just outmoded, it is plainly wrong. These are prejudices that survive from a former era.

Evolution and Theology[81]

It is my belief that the Darwinian way of explaining the evolution of so-called "higher" organic forms from the "lower" ones, instead of rendering theology irrelevant, actually enriches it. Evolutionary science speaks of our familiar relation to all life, on the one hand, and more intimately to our closeness to all mammals and primarily to primates, on the other hand. Theology, then, as John Haught suggests,[82] can embrace the scientific information provided, in this case by evolutionary biology together with the kind of meta-

---

[79] Midgley, 187.

[80] Midgley, 188; words in quotation marks are Midgley's own.

[81] One of my reasons for this section is a comment by Midgley in regard to her interest in reaching those in theological studies: "I'm particularly pleased when someone reports that they are using [my books] to teach in a theological seminary. I've always wanted to reach this public — as well as the scientists;" (from our first letter exchange through electronic mail on February 4, 2005). When I first wrote to her, I expressed my interest in using her philosophical insights into science and particularly in evolution for my own research and reflection as a theologian.

[82] As mentioned before, John Haught, *God after Darwin: A Theology of Evolution* (Boulder, Colo.: Westview, 2000); in this case, his arguments in 73-74.

physics that is necessary to speak about God's complex presence in a complex world. Midgley has called this kind of work a "vast and difficult, but very important subject,"[83] but one that does not fall in her own territory properly speaking.

Even on this point, and without claiming to be the theologian, Midgley has a cautionary tale. According to Midgley, when it comes to the assertion of a personal God within the framework of a scientific view of the world, there we confront some problems. She thinks, for example, that the notion of a personal cosmic will, which is typically found in an anthropomorphic religious creed, is basically hostile to science. There is no place in natural science for it. She thinks moreover that this notion is not consistent with conceptions of order, so vital to our understanding of the universe (as cosmos). It is order, and not a personal arbitrary will, that the human mind seek to penetrate with the tools of science.[84]

All that said, Midgley likewise has some helpful and constructive suggestions for theologians. From what I can tell, her ideas in this regard can be summarized in a few basic points. First, there is her critique of the mechanistic understanding of the world. According to Midgley, this mechanistic view was scientific as well as religious. From a scientific perspective, the combination of the science of mechanics together with the so-called mathematization of the world is the main sense in which the mechanization of the world-picture in early modernity can be understood.[85] There were certainly other factors, social forces like com-

---

[83] Midgley's electronic mail of July 13, 2005 in response to my letter informing her of the kind of work that I was doing in philosophy of religion inspired by her ideas.

[84] Aspects of this discussion are carried in several of Midgley's works, for example, in "Strange Contest: Science versus Religion," 49; and also in *Evolution as a Religion*, 70.

[85] On this point, see Alfred W. Crosby, *The Measure of Reality: Quantification and Western Society, 1250–1600* (reprint ed.;

merce and technology, as well as intellectual movements like those rejecting any appeal to mystery and offering a critique of superstition, that had their own impact on this change towards a new world-picture. But the image of the universe itself as an immense machine, and the natural philosopher (scientist) as having the knowledge of all its parts and the expertise on its workings, had no parallel in previous history.[86]

In the new science, nature was basically deprived of any magical or miraculous phenomena. Nature has a regularity governed by law, a law that was expressed in mathematical exactness and simplicity. The measure of things, the measure of "reality," became a very important pursuit for the natural philosopher.

From a religious perspective, in a mechanical universe, God became the ultimate designer. However, once God is drawn out of the picture, we are left with a vacuum, soon to be filled with something or someone else. In any case, the machine imagery survives.[87]

In a sense, natural selection in evolutionary theory has been forced to fill this need for a designer of sorts. This development was probably beyond Darwin's own expectations. We could speculate about Darwin's ultimate motives, but the evidence is not clear on this matter. If anything, the evidence from Darwin himself is at most ambiguous in this regard. But natural selection is a mechanism, supposedly a blind one. At best it is a metaphor for a complex process of interaction between the development and diversity of organic life in its relation to the environment. Thus it does

---

Cambridge: Cambridge UP, 1997), 238–239. See also Richard Westfall, *The Construction of Modern Science: Mechanism and Mechanics* (reprint ed.; Cambridge: Cambridge UP, 1995), 1–2.

[86] See Westfall, 30–31.

[87] Midgley, *The Myths We Live By*, 118.

not truly inspire people's imaginations.[88] Natural selection falls short as an answer to people's deepest questions about life's goals and meaning. The crux of the matter remains that humans want to know, not just about natural means or provisional ends, but about destiny—teleological concerns surface time and again. It is not enough for science to deny the meaningfulness of teleological questions; human beings still want to know about life's sense and their own place in the scheme of things.

The vacuum left by God in modern culture is to be filled now by biotechnology instead. Mechanistic thinking still calls for a designer. But the designer is now a person, a scientist and technician, who confidently proposes ways of designing our future. We seem to be looking to remake ourselves by improving our image and capabilities. When the public worries about these developments, wondering about the why and what for and for whom these biotechnological developments are, the public is basically expressing its uneasiness in general terms. So, they make use of whatever language is available to them—for example, faith language—in order to point out that someone somewhere is trying to "play God."

> So it emerges that members of the public who complain that biotechnological projects involve *playing God* have in fact understood this claim correctly. That phrase, which defenders of the projects dismiss as mere mumbo jumbo, is actually a quite exact term for the sort of claim to omniscience and omnipotence on these matters that is being put forward.[89]

Maybe it is the very conception of the world and of ourselves as machines, and therefore the need for a designer, that has to change.

---

[88] Midgley, 118.
[89] Midgley, 118; emphasis is the author's.

Second, there is Midgley's critique of anthropocentrism, or human centeredness. This is her way of calling our attention to the fact that we easily forget that we are just one species among others. We have no warrant to use ideas about human uniqueness as a cover for our general destruction of other creatures and their environments.[90] According to Midgley, there is an extraordinary over-confidence in our powers that cannot be based either in true religion or normal science. There are instances, however, of a pseudo-religion that speaks of humans as the crown of the created order, and as exalted in all respects, from intellect and creativity to the nature of our emotions. There is likewise a pseudo-science, or scientism, which seems to thrive on prophecy and superstition[91] about our alleged increasing powers over nature, as if humans were not part of nature in the first place. Scientism is blind to a reasonable measure of our capacities, especially for the powers of destruction, the ones that we have failed to keep under control. It tends to be blind also to the by-products of human scientific and technological progression.

Midgley is adamant in saying that we are not, as currently believed in many circles, the "point of direction" of the evolutionary process. To think like that is for her evidence of a sentiment of human centeredness that only serves to distort our understanding of evolution. We are neither in charge of defining what the world necessarily is, nor where it needs to go, nor what it ought to be. That would be very pretentious, an arbitrary exercise in human intellect-centered purpose.[92] In theological language, we say that the world is what God wills it to be. Moreover, we affirm that the world is God's, and for God's purposes, and not ours.

---

[90] See, Mary Midgley, *Evolution as a Religion: Strange Hopes and Stranger Fears* (revised ed.; London: Routledge, 2002), 74.

[91] Midgley, 75.

[92] Midgley, 79.

For Midgley, anthropocentrism has been reinforced since the Enlightenment by the idea of the social contract, the "social contract myth," as she calls it. As originally envisioned by its proponents (Rousseau and others), it was a way to assert basic human goodness in a state of nature. It was also a call for a return to our natural origins, especially in the midst of a civilization obsessed with the advancement of the human race through science and technology. There was simultaneously an early reaction against mechanistic science and its perceived mistreatment and devaluation of nature.[93] However, social contract thinking still assumed that what mattered was the human; therefore, there was no acknowledgment of responsibility for others of nature's inhabitants. There was no place or even consideration for non-human animals or for the rest of nature.[94] Everything was considered exclusively from the perspective of human needs. Ethics was clearly confined to human beings. No duties or responsibilities except those to other humans were ever considered.[95]

We are still carrying with us a lot of this cultural baggage. There is need for a change in attitudes toward nature. We seem to have made a virtue of anthropocentrism, the belief that we humans determine the proper value of everything else. Thus I agree with Midgley that, "we profoundly need to get *rid* of something. We need to get rid of the notion that all natural things are valueless in themselves, merely pretty extras, expendable, either secondary to human purposes or actually pernicious."[96] For Midgley, moreover, social contract thinking, which is a product of Enlightenment reasoning, is definitely anthropocentric: it has no place for the non-human world. Moreover, after a while, God was not

---

[93] Midgley, *The Myths We Live By*, 171.
[94] Midgley, 172.
[95] Midgley, 173.
[96] Midgley, 174; emphasis is the author's.

needed in the picture either. Enlightenment anthropocentrism is all about human rational capabilities and self-sufficiency.

The third and final point in our exposition is the importance and coherence of Midgley's "down-to-earth" epistemology. She accepts that what she has developed can be called an "evolutionary epistemology." However, and at the same time, she wants to stay clear from Karl Popper's (1902–1994) understanding, which she deems "highly abstract."[97]

Midgley mentions Popper in this context just because Popper is normally credited with having initiated an early strand of this kind of epistemology by his analysis of the Darwinian theory of evolution. Early in his philosophical career, Popper had criticized Darwinism because it does not stand the validity test of real science: falsifiability.[98] Popper calls Darwinism a "metaphysical research programme," something that is not testable in the same manner as other scientific theories.[99] He meant to say that its main propositions do not stand a rigorist logical analysis and, therefore, lack the coherence of other forms of scientific knowledge. A principle like "natural selection" is basically tautological, because it is circular and thus do not add any information beyond that already included in the definition. For example, if natural selection is guided by the "survival of the fittest," then it definitively begs the question. It is like saying that those who actually survive are the fittest, so those capable of surviving are precisely the fittest, too.

---

[97] This is Midgley in an electronic mail note on February 28, 2006.

[98] A form of verification for a scientific theory by looking for instances that may falsify its claims. We had already referred to this principle in chapter one.

[99] First in Karl Popper, *Objective Knowledge: An Evolutionary Approach* (Oxford: Clarendon, 1972); and then in, among other writings, *Unended Quest: An Intellectual Autobiography* (original ed., 1974; reprint ed.; London: Routledge, 1992), 194–210.

Eventually, however, and as he testified in his autobiography, Popper acknowledged the factual character of evolutionary theory, by what he came to call "situational logic," which he defined as a situation of "trial and error elimination."[100] Moreover, he came to realize that the definition of natural selection was more complex than he had thought at first, and that Darwinism has more than one basic explanatory principle for evolution. As a matter of fact, Darwinism is a set of various interrelated principles. Thus, Darwinism is science but in a rather pluralistic sense.

Important as it is to understand the various layers of meaning that evolution has, what interests Midgley mostly is to learn from Darwin's deep respect and reverence for nature and its organic forms. More relevant than the metaphysical uses and abuses of evolution—of which she has much to say anyway—Midgley is taken by Darwin's humble attitude towards all living things and the intricate interconnections among organisms and their environments: every living thing, from the simplest to the most complex creature, matters. For the task of getting to know who we are as humans, there is need to look all around us, to consider life in all its richness and diversity. The task of knowing life, its origin, development, and possibilities for the future is really daunting. But it did not deter Darwin in his pursuit of knowledge and understanding. Darwin never considered the time and energy spent in the study of the most insignificant of living organism a waste. From his five years (1831–1836) aboard the *Beagle* as an amateur naturalist, to the eight years (c1850–1858) of dedicated attention to barnacles, to the almost forty years of his study of earthworms,[101] to plants, birds and animals of many kinds and

---

[100] Popper, *Unended Quest*, 195–196.
[101] See, Midgley, *The Myths We Live By*, 126. Referring to Darwin's amazing attention to detail, Midgley writes: "Darwin tested the response of the worms to every kind of situation, confronting

species, he proceeded to his study and reflection with almost religious dedication.

Moreover, Midgley sees in Darwin a scientist with a concern for wisdom. Even if often hesitantly and at times almost reluctantly, Darwin still managed to say many things about "who we are" and "what our relationship to everything else" is. And he did it the hard way, by detailed and concentrated observation, and by an amazing openness to the world out there. But also by wondering what it all says about our inside.

The Darwinian understanding of evolution touches on different facets of life on earth. As we have seen, it has different meanings for different people. Above all, and what makes many people nervous, it is a "myth of origins:" it speaks forcefully about who we are and where we come from. It truly puts us in perspective, while keeping a humble tone. In Midgley's view, Darwinian evolution situates us down here where we truly belong.

It tells us that the earth is our home.

---

them with all sorts of experience: different kinds of light, heat, smells, vibrations and music, including the bassoon and the grand piano... Moreover, by investigating these things, he made the revolutionary discovery that earthworms, which had till then been considered either insignificant or pestilential, in fact played a central part in recycling vegetation and turning it into usable soil. Without them, this process would be far too slow for other life forms to profit by it."

Bibliography

PRIMARY SOURCES

Books

Midgley, Mary. *Animals and Why They Matter.* Athens, Georgia: The University of Georgia, 1983.
– – – *Beast and Man: The Roots of Human Nature.* 1979. Rev. ed. London: Routledge, 2002.
– – – *Biological and Cultural Evolution.* Kent, Eng.: The Institute for Cultural Research, 1984.
– – – *Can't We Make Moral Judgements?* 1991. New York: St. Martin's, 1993.
– – – *Evolution as a Religion: Strange Hopes and Stranger Fears.* 1985. Rev. ed. London: Routledge, 2002.
– – – *Gaia: The Next Big Idea.* London: Demos, 2001.
– – – *Heart and Mind: The Varieties of Moral Experience.* 1981. Rev. ed. London: Routledge, 2003.
– – – *Science and Poetry.* 2001. London: Routledge, 2002.
– – – *Science as Salvation: A Modern Myth and its Meanings.* 1992. London: Routledge, 1994.
– – – *The Essential Mary Midgley.* Ed. David Midgley. London: Routledge, 2005.
– – – *The Ethical Primate: Humans, Freedom and Morality.* 1994. London: Routledge, 1996.
– – – *The Myths We Live By.* London: Routledge, 2003.
– – – *The Owl of Minerva: A Memoir.* London: Routledge, 2005.
– – – *Utopias, Dolphins and Computers: Problems of Philosophical Plumbing.* 1996. London: Routledge, 2000.
– – – *Wickedness: A Philosophical Essay.* 1984. London: Routledge, 2001.

――― *Wisdom, Information, and Wonder: What is Knowledge for?* 1989. London: Routledge, 1995.
――― and Judith Hughes. *Women's Choices: Philosophical Problems Facing Feminism*. London: Weidenfeld and Nocolson, 1983.
――― ed. *Earthy Realism: The Meaning of Gaia*. Exeter: Imprint Academic, 2007.

What follows is the list of those writings that have not been reproduced in their entirety in any of her published books; those articles that have been fully incorporated in her books are omitted below

Book Chapters

Midgley, Mary. "Darwinism and Ethics." *Medicine and Moral Reasoning*. Ed. K. W. M. Fulford, Grant R. Gillett, and Janet Martin Soskice. Cambridge: Cambridge UP, 1994. 6-18.
――― "Dover Beach: Understanding the Pains of Bereavement." *Oxford Handbook of Religion and Science*. Oxford: Oxford UP; forthcoming, 2006.
――― "Human Nature, Human Variety, Human Freedom." *Being Humans: Anthropological Universality and Particularity in Transdisciplinary Perspectives*. Ed. Neil Roughley. Berlin: Walter de Gruyter, 2000. 47-63.
――― "Intelligence, Wisdom and Folly." *Where Shall Wisdom be Found? Wisdom in the Bible, the Church and the Contemporary World*. Ed. Stephen C. Barton. Edinburgh: T and T Clark, 1999. 185-193.
――― "Keeping Species on Ice." *Beyond the Bars: The Zoo Dilemma*. Ed. Virginia McKenna. Rochester, Vt.: Thorsons, 1987. 55-65.
――― "Memes." *Encyclopedia of Science and Religion*. Vol. 2. Ed. J. Wentzel Vrede van Huyssteen. New York: MacMillan, 2003. 556-557.
――― "Murdoch and Morality." *What Philosophers Think*. Ed. Julian Baggini and Jeremy Stangroom. London: Continuum, 2003. 125-132.
――― "On Being Terrestrial." *Objectivity and Cultural Divergence.* Supplement to *Philosophy* 1984. Ed. S. C. Brown. Cambridge: Cambridge UP, 1984. 79-91.
――― "On Not Being Afraid of Natural Sex Differences." *Feminist Perspectives in Philosophy*. Ed. Morwenna Griffiths and Margaret Whitford. Bloomington: Indiana UP, 1988. 29-41.

― ― ― "Persons and Non-Persons." *In Defence of Animals.* Ed. Peter Singer. Oxford: Basil Blackwell, 1985. 52-62.

― ― ― "Rival fatalisms: The Hollowness of the Sociobiology Debate." *Sociobiology Examined.* Ed. Ashley Montagu. New York: Oxford UP, 1980. 15-38.

― ― ― "Selfish Gene." *Encyclopedia of Science and Religion.* Vol. 2. Ed. J. Wentzel Vrede van Huyssteen. New York: MacMillan, 2003. 795-796.

― ― ― "Strange Contest: Science versus Religion." *The Gospel in Contemporary Culture.* Ed. Hugh Montefiore. London: Mowbray, 1992. 40-57.

― ― ― "The Challenge of Science: Limited Knowledge or New High Priesthood?" *True to this Earth: Global Challenges and Transforming Faith.* Ed. Alan Race and Roger Williamson. Oxford: Oneworld, 1995. 75-84.

― ― ― "The Mixed Community." *Earth Ethics: Environmental Ethics, Animal Rights, and Practical Applications.* Ed. James P. Sterba. Englewood Cliffs: Prentice Hall, 1995. 80-90.

― ― ― "The Need for Wonder." *God for the 21st Century.* Ed. Russell Stannard. Philadelphia: Templeton Foundation, 2000. 186-188.

― ― ― "The Origins of Ethics." *A Companion to Ethics.* Ed. Peter Singer. 1991. Oxford: Blackwell, 1997. 3-13.

― ― ― "The Paradox of Humanism." *James M. Gustafson's Theocentric Ethics: Interpretations and Assessments.* Ed. Harlan R. Beckley and Charles M. Swezey. Macon: Mercer UP, 1988. 187-199.

― ― ― "The Problem of Humbug." *Media Ethics.* Ed. Matthew Kieran. London: Routledge, 1998. 37-48.

― ― ― "The Problem of Living with Wildness." *Wolves and Human Communities.* Ed. Virginia A. Sharpe, Bryan G. Norton, and Strachan Donnelley. Washington, D.C.: Island, 2000. 179-190.

― ― ― "The Reality of Human Wickedness." *Applied Ethics and Ethical Theory.* Vol. 1: *Ethics in a Changing World.* Ed. David M. Rosenthal and Fadlou Shehadi. Salt Lake City: University of Utah, 1988. 306-321.

― ― ― "The Religion of Evolution." *Darwinism and Divinity: Essays on Evolution and Religious Belief.* Oxford: Blackwell, 1985. 155-180.

― ― ― "The Significance of Species." *The Animal Rights/Environmental Ethics Debate.* Ed. Eugene C.

Hargrove. Albany, NY: State University of New York, 1992. 121-136.

——— "Towards a More Humane View of the Beast." *The Environment in Question: Ethics and Global Issues.* Ed. David Cooper. London: Routledge, 1992. 28-36.

——— "Towards a New Understanding of Human Nature: The Limits of Individualism." *How Humans Adapt: A Biocultural Odyssey.* Ed. Donald J. Ortner. Washington, D.C.: Smithsonian Institution, 1983. 517-533.

——— "Why Smartness is not Enough." *Rethinking the Curriculum: Toward and Integrated, Interdisciplinary College Education.* Ed. Mary E. Clark and Sandra A. Wawrytko. New York: Greenwood, 1990. 39-52.

Journal Articles

Midgley, Mary. "A Plague On Both Their Houses." *Philosophy Now* 64 (2007): 26-27.

——— "Alchemy Revived." *The Hasting Center Report* 30/2 (2000):

——— "Being Objective." *Nature* 410 (2001): 753.

——— "Brutality and Sentimentality." *Philosophy* 54 (1979): 385-389.

——— "Can Education Be Moral?" *Res Publica: A Journal of Legal and Social Philosophy* 2 (1996): 77-85.

——— "Can Science Save its Soul?" *New Scientist* 135 (1992): 24-27.

——— "Determinism, Omniscience, and the Multiplicity of Explanations." *Behavioral and Brain Sciences* 22 (1999): 900-901.

——— "Evolution as Religion: A Comparison of Prophecies." *Zygon* 22 (1987): 179-194.

——— "Gene-Juggling." *Philosophy* 54 (1979): 439-458.

——— "Heaven and Earth: An Awkward History." *Philosophy Now* 34 (2001/2002): 18-21.

——— "Human Ideas and Human Needs." *Philosophy* 58 (1983): 89-94.

——— "Individualism and the Concept of Gaia." *Review of International Studies* 26 (2000): 29-44.

——— "Is the Biosphere a Luxury?" *The Hastings Center Report* 22 (1992): 7-12.

——— "Madness in Method." *Times Higher Education Supplement* (March 28 1997): 18-19.

― ― ― "More about Reason, Commitment and Social Anthropology." *Philosophy* 53 (1978): 401-403.
― ― ― "Must Good Causes Compete?" *Cambridge Quarterly of Healthcare Ethics* 2 (1993): 133.
― ― ― "Pluralism; The Many-Maps Model." *Philosophy Now* 35 (2002): 10-11.
― ― ― "Practical Solutions." *Hastings Center Report* 19 (1989): 44-45.
― ― ― "Reductivism, Fatalism and Sociobiology." *Journal of Applied Philosophy* 1 (1984): 107-114.
― ― ― "Selfish Genes and Social Darwinism." *Philosophy* 58 (1983): 365-377.
― ― ― "Skimpole Unmasked." *History of the Human Sciences* 10 (1997): 92-96.
― ― ― "Sociobiology." *Journal of Medical Ethics* 10 (1984): 158-160.
― ― ― "Sorting Out the Zeitgeist: The Moral Philosophy of Iris Murdoch." *The Philosopher* 86 (1998).
― ― ― "Souls, Minds, Bodies and Planets (part 1)." *Philosophy Now* 47 (2004): 33-35.
― ― ― "Souls, Minds, Bodies and Planets (part 2)." *Philosophy Now* 48 (2004): 10-12.
― ― ― "The All Female Number." *Philosophy* 54 (1979): 552-554.
― ― ― "The Flight from Blame." *Philosophy* 62 (1987): 271-291.
― ― ― "The Four-Leggeds, the Two-Leggeds, and the Wingeds." *Society and Animals* 1/1 (1993).
― ― ― "The Lack of Gap between Fact and Value." *Proceedings of the Aristotelian Society* 54 (1980): 207-224.
― ― ― "The Morality of Creature Comfort." *Times Higher Education Supplement* 1288 (July 11 1997): 20-21.
― ― ― "The Notion of Instinct." *Cornell Review* 7 (1979).
― ― ― "Visions and Values." *Resurgence* 228 (2005).
― ― ― "What Do We Mean By Security?" *Philosophy Now* 61 (2007): 12-15.
― ― ― "Wickedness." *Encyclopedia of Ethics*. 2nd ed. Ed. Lawrence C. Becker and Charlotte B. Becker. Routledge, 2001. Routledge Reference Resources Online. Taylor and Francis Publishing Group. <http://www.reference.routledge.com>
― ― ― "Zombies and the Turing Test." *Journal of Consciousness Studies* 2 (1995): 351-352.

— — — and Stephen R. L. Clark. "The Absence of a Gap between Facts and Values." *The Aristotelian Society*, Supplementary Volume 54 (1980): 207-223.

Other

Midgley, Mary et al. "Environment and Humanity: Friends or Foes?" Conversations with the Archbishop, St. Paul's Institute of St. Paul's Cathedral in London, September 21, 2004 <http://www.stpauls.co.uk>

— — — and Nick Spencer. *Discussing Darwin: An Extended Interview with Mary Midgley.* London: Theos, 2009. Also available at Theos: The Public Theology Think Tank <http://www.theosthinktank.co.uk>

SECONDARY SOURCES

Agamben, Giorgio. *The Open: Man and Animal.* Stanford, Cal.: Stanford UP, 2004.

Alcock, John. *The Triumph of Sociobiology.* Oxford: Oxford UP, 2001.

Almog, Joseph. *What Am I? Descartes and the Mind-Body Problem.* Oxford: Oxford UP, 2002.

Amigoni, David, and Jeff Wallace, eds. *Charles Darwin's 'The Origin of Species': New Interdisciplinary Essays.* Manchester, U.K.: Manchester UP, 1995.

Antonaccio, Maria. *Picturing the Human: The Moral Thought of Iris Murdoch.* New York: Oxford UP, 2000.

Appleyard, Bryan. *Understanding the Present: Science and the Soul of Modern Man.* New York: Doubleday, 1992.

Atran, Scott. *In Gods We Trust: The Evolutionary Landscape of Religion.* Oxford: Oxford UP, 2002.

Ayer, A. J. *Language, Truth, and Logic.* London: Gollanz, 1936.

Baeyer, Hans Christian von. *Information: The New Language of Science.* Cambridge, Mass.: Harvard UP, 2004.

Baldi, Pierre. *The Shattered Self: The End of Natural Evolution.* 2001. Cambridge, Mass.: MIT, 2002.

Barbour, Ian G. *Myths, Models and Paradigms: A Comparative Study in Science and Religion.* New York: Harper and Row, 1974.

— — — *Religion and Science: Historical and Contemporary Issues.* 1990. Revised and expanded ed. New York: Harper San Francisco, 1997.

Barlow, Connie, ed. *Evolution Extended: Biological Debates on the Meaning of Life.* Cambridge, Mass.: MIT, 1994.

Barlow, Nora, ed. *The Autobiography of Charles Darwin (1809-1882)*. Complete ed. 1958. Reprint ed. New York: Norton, 1993.

Barr, Stephen. "The Devil's Chaplain Confounded." *First Things* 145 (2004): 25-30.

Blackburn, Simon. *Think: A Compelling Introduction to Philosophy*. Oxford: Oxford UP, 1999.

Blackmore, Susan. *The Meme Machine*. 1999. Oxford: Oxford UP, 2000.

Bove, Cheryl K. *Understanding Iris Murdoch*. Columbia: University of South Carolina, 1993.

Bowker, John. *Is God a Virus? Genes, Culture and Religion*. London: SPCK, 1995.

Boyer, Pascal. *Religion Explained: The Evolutionary Origins of Religious Thought*. New York: Basic Books, 2001.

Brooke, John H. *Science and Religion: Some Historical Perspectives*. 1991. Cambridge: Cambridge UP, 1993.

— — — -, Margaret J. Osler, and Jitse M. van der Meer, eds. *Science in Theistic Contexts: Cognitive Dimensions*. Chicago: Osiris, 2001.

Brown, Andrew. *The Darwin Wars: The Scientific Battle for the Soul of Man*. London: Simon and Schuster, 1999.

Buller, David J. *Adapting Minds: Evolutionary Psychology and the Persistent Quest for Human Nature*. Cambridge, Mass.: MIT, 2005.

Caudill, Edward. *Darwinian Myths: The Legends and Misuses of a Theory*. Knoxville: The University of Tennessee, 1997.

Clayton, Philip. "Biology Meets Theology." *Christian Century* 117 (2000): 61-64.

— — — *Mind and Emergence: From Quantum to Consciousness*. Oxford: Oxford UP, 2004.

— — — and Jeffrey Schloss, eds. *Evolution and Ethics: Human Morality in Biological and Religious Perspective*. Grand Rapids, Mich.: Eerdmans, 2004.

Conradi, Peter J. *Iris: The Life of Iris Murdoch*. New York: Norton, 2001.

Cornwell, John, ed. *Nature's Imagination: The Frontiers of Scientific Vision*. Oxford: Oxford UP, 1995.

Crick, Francis. *What Mad Pursuit*. London: Penguin, 1989.

Cronk, Lee. *That Complex Whole: Culture and the Evolution of Human Behavior*. Boulder, Colo.: Westview, 1999.

Crosby, Alfred W. *The Measure of Reality: Quantification and Western Society, 1250-1600.* Reprint ed. Cambridge: Cambridge UP, 1997.

Damasio, Antonio. *Descartes' Error: Emotion, Reason, and the Human Brain.* 1994. New York: Quill, 2000.

Darwin, Charles. *The Descent of Man, and Selection in Relation to Sex.* 1871. 2nd ed. 1879. London: Penguin, 2004.

— — — *The Expression of Emotions in Man and Animal.* 1872. 2nd ed. 1889. Oxford: Oxford UP, 1998.

— — — *The Origin of Species.* 1859. New York: The Modern Library, 1998.

Darwin, Francis, ed. *The Life and Letters of Charles Darwin.* 1887. Reprint ed. New York: Appleton, 1905.

Davis, Merryl Wyn. *Darwin and Fundamentalism.* Cambridge: Icon Books, 2000.

Dawkins, Richard. *A Devil's Chaplain: Reflections on Hope, Lies, Science and Love.* Boston: Houghton Mifflin, 2003.

— — — "In Defense of Selfish Genes" [A Response to Midgley's "Gene-Juggling"]. *Philosophy* 56 (1981): 556-573.

— — — "Is Science a Religion?" *The Humanist: A Magazine of Critical Inquiry and Social Concern* (January/February, 1997). The Humanist Online <http://thehumanist.org/humanist/articles/dawkins.html>

— — — *River Out of Eden: A Darwinian View of Life.* New York: Basic Books, 1995.

— — — *Unweaving the Rainbow: Science, Delusion and the Appetite for Wonder.* 1998. Boston: Houghton Mifflin, 2000.

— — — *The Blind Watchmaker: Why the Evidence of Evolution Reveals a Universe without Design.* 1986. New York: Norton, 1996.

— — — *The Extended Phenotype: The Long Reach of the Gene.* 1982. Oxford: Oxford UP, 1999.

— — — *The Selfish Gene.* 1976. Oxford: Oxford UP, 1999.

— — — "Viruses of the Mind." *Dennett and his Critics: Demystifying Mind.* Ed. Bo Dahlbom. Reprint ed. Oxford: Blackwell, 1994. 13-27.

Day, William. *Genesis on Planet Earth: The Search for Life's Beginning.* 2nd ed. New Haven: Yale UP, 1984.

Dennen, Johan M. G. van der, et al., eds. *The Darwinian Heritage and Sociobiology.* Westport, Conn.: Praeger, 1999.

Dennett, Daniel C. *Breaking the Spell: Religion as a Natural Phenomenon.* New York: Viking, 2006.

— — — *Darwin's Dangerous Idea: Evolution and the Meanings of Life*. New York: Simon and Schuster, 1995.

Derksen, Ton, ed. *The Promise of Evolutionary Epistemology*. Tilburg: Tilburg, The Netherlands: Tilburg UP, 1998.

Desmond, Adrian and James Moore. *Darwin: The Life of a Tormented Evolutionist*. 1991. New York: Norton, 1994.

Distin, Kate. *The Selfish Meme: A Critical Reassessment*. Cambridge: Cambridge UP, 2005.

Dobzhansky, Theodosius G. *The Biology of Ultimate Concern*. New York: The New American Library, 1967.

Dowe, Phil. *Galileo, Darwin, Hawking: The Interplay of Science, Reason, and Religion*. Grand rapids, Mich.: Eerdmans, 2005.

Dupré, John. *Darwin's Legacy: What Evolution Means Today*. Oxford: Oxford UP, 2004.

— — — *Humans and Other Animals*. Oxford: Clarendon, 2002.

Duran, Jane. *Philosophies of Science/Feminist Theories*. Boulder, Colo.: Westview, 1998.

Durant, John, ed. *Darwinism and Divinity: Essays on Evolution and Religious Belief*. Oxford: Blackwell, 1985.

Eldredge, Niles. *Discovering the Tree of Life*. New York: Norton, 2005.

— — — *Reinventing Darwin*. New York: Wiley, 1995.

— — — *Why We Do It: Rethinking Sex and the Selfish Gene*. New York: Norton, 2004.

Farber, Paul Lawrence. *The Temptations of Evolutionary Ethics*. Berkeley: University of California, 1994.

Fernández-Armesto, Felipe. *Humankind: A Brief History*. Oxford: Oxford UP, 2004.

Fichman, Martin. *Evolutionary Theory and Victorian Culture*. Amherst, N.Y.: Humanity Books, 2002.

Flew, Anthony. *A Dictionary of Philosophy*. 1979. 2nd rev. ed. New York: St. Martin's, 1984.

Fukuyama, Francis. *Our Posthuman Future: Consequences of the Biotechnology Revolution*. New York: Picador, 2002.

Fuller, Steve. *Kuhn vs. Popper: The Struggle for the Soul of Science*. New York: Columbia UP, 2004.

Gebara, Ivone. *Longing for Running Water: Ecofeminism and Liberation*. Trans. David Molineaux. Minneapolis: Fortress Press, 1999.

Gerhart, Mary and Allan Russell. *Metaphoric Process: The Creation of Scientific and Religious Understanding.* Fort Worth: Texas Christian University, 1984.

— — — *New Maps for Old: Explorations in Science and Religion.* New York: Continuum, 2001.

Giberson, Karl W. and Donald A. Yerxa. *Species of Origin: America's Search for a Creation Story.* Lanham, Maryland: Rowman and Littlefield, 2002.

Goodenough, Ursula. *The Sacred Depths of Nature.* Oxford: Oxford UP, 1998.

Gould, Stephen J. "Darwinian Fundamentalism." *The New York Review of Books* 44/10 (June 12, 1997): 34-37.

— — — *Ever Since Darwin: Reflections on Natural History.* 1973. New York: Norton, 1992.

— — — "Evolution: The Pleasures of Pluralism." *New York Review of Books* 44/11 (June 26, 1997): 47-52.

— — — *Rocks of Ages: Science and Religion in the Fullness of Life.* New York: Ballantine, 1999.

— — — *The Structure of Evolutionary Theory.* Cambridge, Mass.: Belknap/Harvard UP, 2002.

— — — and Richard C. Lewontin, "The Spandrels of San Marco and the Panglossian Paradigm: A Critique of the Adaptationist Programme." *Proceedings of the Royal Society of London.* Series B: *Biological Sciences* 205 (1979): 581-598.

Graham, Gordon. *Genes: A Philosophical Inquiry.* London: Routledge, 2004.

Gray, John. *Straw Dogs: Thoughts on Humans and Other Animals.* London: Granta, 2002.

Greene, Brian. *The Elegant Universe: Superstrings, Hidden Dimensions, and the Quest for the Ultimate Theory.* New York: Vintage, 1999.

Greene, John C. *Debating Darwin: Adventures of a Scholar.* Claremont, Cal.: Regina Books, 1999.

Griffiths, Sian, ed. *Beyond the Glass Ceiling: Forty Women Whose Ideas Shape the Modern World.* Manchester: Manchester University, 1996.

Gustafson, James M. *A Sense of the Divine: The Natural Environment from a Theocentric Perspective.* Cleveland, Ohio: Pilgrim, 1997.

Haack, Susan. *Defending Science: Within Reason: Beyond Scientism and Cynicism.* Amherst, N.Y.: Prometheus, 2003.

Hall, Stephen S. "Darwin's Rottweiler." *Discover* 26/9 (Sept. 2005): 51-56.

Harding, Sandra. *The Science Question in Feminism*. Ithaca: Cornell UP, 1986.

Harth, Erich. *Dawn of the Millennium: Beyond Evolution and Culture*. Boston: Little, Brown and Co., 1990.

Haught, John F. *Deeper than Darwin: The Prospect of Religion in the Age of Evolution*. Cambridge, Mass.: Westview, 2003.

– – – *God after Darwin*. Cambridge, Mass.: Westview, 1999.

– – – "The Darwinian Universe: Isn't there Room for God? *Commonweal* 129/2 (2002): 12-18.

– – – *Science and Religion: From Conflict to Conversation*. New York: Paulist, 1995.

Herzfeld, Noreen L. *In Our Image: Artificial Intelligence and the Human Spirit*. Minneapolis: Fortress, 2002.

Hodge, Jonathan, and Gregory Radick, eds. *The Cambridge Companion to Darwin*. Cambridge: Cambridge UP, 2003.

Horgan, John. *The End of Science: Facing the Limits of Knowledge in the Twilight of the Scientific Age*. Reading, Mass.: Addison Wesley, 1996.

Hösle, Vittorio, and Christian Illies, eds. *Darwinism and Philosophy*. Notre Dame, Ind.: University of Notre Dame, 2005.

Hursthouse, Rosalind. *Ethics, Humans and Other Animals: An Introduction with Readings*. London: Routledge, 2000.

Jablonka, Eva, and Marion J. Lamb. *Evolution in Four Dimensions: Genetic, Epigenetic, Behavioral and Symbolic Variation*. Cambridge, Mass.: MIT, 2005.

James, William. *The Varieties of Religious Experience*. 1902. Reprint ed. New York: Barnes and Noble, 2004.

Jones, Steve. *Darwin's Ghost: The 'Origen of Species' Updated*. 1999. New York: Ballantine, 2000.

Keller, Evelyn Fox. *The Century of the Gene*. Cambridge, Mass.: Harvard UP, 2000.

– – – and Elisabeth A. Lloyd, eds. *Keywords in Evolutionary Biology*. Cambridge, Mass.: Harvard UP, 1992.

– – – and Helen E. Longino, eds. *Feminism and Science*. Oxford: Oxford UP, 1996.

Keynes, Randal. *Darwin, His Daughter, and Human Evolution*. New York: Riverhead, 2001.

Kuhn, Thomas S. *The Structure of Scientific Revolutions*. 1962. 3rd ed. Chicago: The University of Chicago, 1996.

Kuper, Adam. *The Chosen Primate: Human Nature and Cultural Diversity*. Cambridge, Mass.: Harvard UP, 1994.

Larson, Edward J. *Evolution: The Remarkable History of a Scientific Theory*. New York: The Modern Library, 2004.

Lewis, John, ed. *Beyond Chance and Necessity: A Critical Inquiry into Professor Jacques Monod's Chance and Necessity*. London: Garnstone, 1974.

Lewontin, R. C. *Biology as Ideology: The Doctrine of DNA*. New York: Harper Perennial, 1991.

Lindberg, David C., and Ronald L. Numbers, eds. *God and Nature: Historical Essays on the Encounter between Christianity and Science*. Berkeley: University of California, 1986.

Livingstone, David N. *Darwin's Forgotten Defenders: The Encounter between Evangelical Theology and Evolutionary Thought*. Grand Rapids, Mich.: Eerdmans and Edinburgh: Scottish Academic, 1987.

Lomolino, Mark et al. *Biogeography*. 3rd ed. Sunderland, Mass.: Sinauer, 2005.

Lorenz, Konrad. *On Aggression*. Trans. Marjorie Kerr Wilson. New York: Harcourt, Brace and World, 1963.

Lovelock, James E. *Gaia: A New Look at Life on Earth*. Oxford: Oxford UP, 1979.

――― *Homage to Gaia: The Life of an Independent Scientist*. 2000. Oxford: Oxford UP, 2001.

――― *The Ages of Gaia: A Biography of Our Living Earth*. New York: Norton, 1988.

Lloyd, Elisabeth A. *The Case of the Female Orgasm: Bias in the Science of Evolution*. Cambridge, Mass.: Harvard UP, 2005.

Mackie, J. L. "Genes and Egoism." *Philosophy* 56 (1981): 553-555.

――― "The Law of the Jungle: Moral Alternatives and Principles of Evolution." *Philosophy* 53 (1978): 455-464.

Margulis, Lynn. *Symbiotic Planet: A New Look at Evolution*. New York: Basic Books, 1998.

McGrath, Alister. *Dawkins' God: Genes, Memes, and the Meaning of Life*. Oxford: Blackwell, 2005.

――― *The Reenchantment of Nature: The Denial of Religion and the Ecological Crisis*. New York: Doubleday, 2002.

Martin, Elizabeth, ed. *A Concise Dictionary of Biology*. 2nd ed. Oxford: Oxford UP, 1990.

Maxwell, Nicholas. *Is Science Neurotic?* London: Imperial College, 2004.

Bibliography 229

Maynard Smith, John. *Shaping Life: Genes, Embryos and Evolution.* New Haven: Yale UP, 1998.

Mayr, Ernst. *One Long Argument: Charles Darwin and the Genesis of Modern Evolutionary Thought.* Cambridge, Mass.: Harvard UP, 1991.

— — — "Speciational Evolution or Punctuated Equilibria." *The Dynamics of Evolution.* New York: Cornell UP, 1992. 21-48.

— — — *What Evolution Is.* New York: Basic Books, 2001. Miller, James B., ed. *An Evolving Dialogue: Theological and Scientific Perspectives on Evolution.* Harrisburg: Trinity, 2001.

Miller, Kenneth. *Finding Darwin's God: A Scientist's Search for Common Ground between God and Evolution.* 1999. New York: Perennial, 2002.

Moltmann, Jürgen. *Science and Wisdom.* Trans. Margaret Kohl. Minneapolis: Fortress, 2003.

Monod, Jacques. *Chance and Necessity: An Essay on the Philosophy of Modern Biology.* Trans. Austryn Wainhouse. New York: Knopf, 1971.

Montagu, Ashley, ed. *Sociobiology Examined.* New York: Oxford UP, 1980.

Moore, G. E. *Principia Ethica.* 1903. Reprint ed. New York: Barnes and Noble, 2005.

Morowitz, Harold J. *The Emergence of Everything: How the World Became Complex.* Oxford: Oxford UP, 2002.

Morris, Richard. *The Evolutionists: The Struggle for Darwin's Soul.* New York: Henry Holt, 2001.

Morris, Simon Conway. *Life's Solution: Inevitable Humans in a Lonely Universe.* Cambridge: Cambridge UP, 2003.

Murdoch, Iris. *Metaphysics as a Guide to Morals.* 1992. London: Penguin Books, 1993.

— — — *Existentialists and Mystics.* London: Penguin Books, 1998.

— — — *The Sovereignty of Good.* 1970. London: Routledge, 2003.

Naam, Ramez. *More Than Human: Embracing the Promise of Biological Enhancement.* New York: Broadway, 2005.

Nagel, Thomas. *What Does It All Mean? A Very Short Introduction to Philosophy.* New York: Oxford UP, 1987.

Nasr, Seyyed Hossein. *Man and Nature: The Spiritual Crisis of Modern Man.* 1968. Chicago: ABC International Group, 1997.

Newton, Isaac. *Mathematical Principles of Natural Philosophy.* 2 vols. 1689. Trans. Andrew Motte, 1729. Rev. Florian Cajori. 6th printing. Berkeley, Cal.: University of California, 1966.

Noble, David F. *The Religion of Technology: The Divinity of Man and the Spirit of Invention.* 1997. New York: Penguin, 1999.

O'Connor, Patricia J. *To Love the Good: The Moral Philosophy of Iris Murdoch.* New York: Peter Lang, 1996.

O'hear, Anthony. *Beyond Evolution: Human Nature and the Limits of Evolutionary Explanation.* Oxford: Clarendon, 1997.

Okasha, Samir. *Philosophy of Science: A Very Short Introduction.* Oxford: Oxford UP, 2002.

Olafson, Frederick A. *Naturalism and the Human Condition: Against Scientism.* London: Routledge, 2001.

Palumbi, Stephen R. *The Evolution Explosion: How Humans Cause Rapid Evolutionary Change.* New York: Norton, 2001.

Pasternak, Charles. *Quest: The Essence of Humanity.* West Sussex, Eng.: Wiley, 2003.

Peacocke, Arthur. *God and the New Biology.* London: J. M. Dent and Sons, 1986.

Phillips, Adam. *Darwin's Worms.* New York: Basic, 2000.

Phipps, William E. *Darwin's Religious Odyssey.* Harrisburg: Trinity, 2002.

Pigliucci, Massimo. "Expanding Evolution." *Nature* 435 (2005): 565-566.

Polkinghorne, John. *Beyond Science: The Wider Human Context.* Cambridge: Cambridge UP, 1996.

Popper, Karl R. *The Logic of Scientific Discovery.* 1959. Reprint ed. London: Routledge, 1992.

——— *Unended Quest: An Intellectual Autobiography.* 1974. New ed. London: Routledge, 2002.

Quammen, David. "Was Darwin Wrong?" *National Geographic Magazine.* National Geographic.Com. Online Extra. November 2004. <http://magma.nationalgeographic.com/ngm/0411/feature1/index.html>

Rachels, James. *Created from Animals: The Moral Implications of Darwinism.* Oxford: Oxford UP, 1991.

Radcliffe Richards, Janet. *Human Nature after Darwin: A Philosophical Introduction.* London: Routledge, 2000.

Rahner, Karl. *Theological Investigations.* Vol. XIII. Trans. David Bourke. New York: Seabury, 1975.

Ratzsch, Del. *Science and Its Limits: The Natural Sciences in Christian Perspective.* 1986. Downers Grove, Ill.: Inter-Varsity, 2000.

Rée, Jonathan, Michael Ayers, and Adam Westoby. *Philosophy and Its Past.* Atlantic Highlands, N.J.: Humanities P, 1978.

Richerson, Peter J, and Robert Boyd. *Not By Genes Alone: How Culture Transformed Human Evolution*. Chicago: The University of Chicago, 2005.

Ridley, Matt. *Nature via Nurture: Genes, Experience, and What Makes Us Human*. New York: Harper Collins, 2003.

– – – *The Origins of Virtue: Human Instincts and the Evolution of Cooperation*. 1996. New York: Penguin, 1998.

Rolston, Holmes. *Genes, Genesis and God: Values and their Origins in Natural and Human History*. Cambridge: Cambridge UP, 1999.

Rose, Hilary and Steven Rose, eds. *Alas, Poor Darwin: Arguments against Evolutionary Psychology*. New York: Harmony, 2000.

Rose, Steven. *Lifelines: Life Beyond the Gene*. 1997. Oxford: Oxford UP, 1998.

Rothschild, Richard C. *The Emerging Religion of Science*. New York: Praeger, 1989.

Rue, Loyal. *Everybody's Story: Wising Up to the Epic of Evolution*. Albany, N.Y.: SUNY, 2000.

Ruse, Michael. *Can a Darwinian be a Christian? The Relationship between Science and Religion*. Cambridge: Cambridge UP, 2001.

– – – *Darwin and Design: Does Evolution Have a Purpose?* Cambridge, Mass.: Harvard UP, 2003.

– – – *Mystery of Mysteries: Is Evolution a Social Construction?* Cambridge, Mass.: Harvard UP, 1999.

– – – *Taking Darwin Seriously: A Naturalistic Approach to Philosophy*. Oxford: Blackwell, 1986.

– – – *The Evolution-Creation Struggle*. Cambridge, Mass.: Harvard UP, 2005.

Segerstråle, Ullica. *Defenders of the Truth: The Battle for Science in the Sociobiology Debate and Beyond*. Oxford: Oxford UP, 2000.

Sexton, Ed. *Dawkins and the Selfish Gene*. Cambridge: Icon Books, 2001.

Shanahan, Timothy. *The Evolution of Darwinism: Selection, Adaptation, and Progress in Evolutionary Biology*. Cambridge: Cambridge UP, 2004.

Singer, Peter. *A Darwinian Left: Politics, Evolution and Cooperation*. New Haven: Yale UP, 2000.

Solberg, Mary M. *Compelling Knowledge: A Feminist Proposal for an Epistemology of the Cross*. New York: SUNY, 1997.

Sorell, Tom, ed. *The Rise of Modern Philosophy: The Tension between the New and Traditional Philosophies from Machiavelli to Leibniz.* Oxford: Clarendon, 1995.

Sperber, Dan. *Explaining Culture: A Naturalistic Approach.* Oxford: Blackwell, 1996.

Stanford, Craig. *Significant Others: The Ape-Human Continuum and the Quest for Human Nature.* New York: Basic Books, 2001.

Stanovich, Keith E. *The Robot's Rebellion: Finding Meaning in the Age of Darwin.* Chicago: The University of Chicago, 2004.

Stenmark, Mikael. "Contemporary Darwinism and Religion." *Darwinian Heresies.* Ed. Abigail Lustig et al. Cambridge: Cambridge UP, 2004. 173-191.

— — — *Scientism: Science, Ethics and Religion.* Aldershot, Eng.: Ashgate, 2001.

Sterelny, Kim. *Dawkins vs. Gould: Survival of the Fittest.* Cambridge: Icon Books, 2001.

Stove, David. "So You Think You Are A Darwinian?" *Philosophy* 69 (1994): 267-277.

Tattersall, Ian. *The Monkey in the Mirror: Essays on the Science of What Makes Us Human.* San Diego: Harcourt, 2002.

Theissen, Gerd. *Biblical faith: An Evolutionary Approach.* Trans. John Bowden. Philadelphia: Fortress, 1985.

Thomson, Keith. *Before Darwin: Reconciling God and Nature.* New Haven: Yale UP, 2005.

Turney, Jon. *Lovelock and Gaia: Signs of Life.* New York: Columbia UP, 2003.

Van Iersel, Bas, Christoph Theobald, and Hermann Häring, eds. *Evolution and Faith.* Continuum 2000/1. London: SCM, 2000.

Verkamp, Bernard J. *The Evolution of Religion: A Re-Examination.* Scranton, Penna.: University of Scranton, 1995.

Waal, Frans de. *The Ape and the Sushi Master: Cultural Reflections of a Primatologist.* New York: Basic Books, 2001.

Westfall, Richard. *The Construction of Modern Science: Mechanism and Mechanics.* Reprint ed. Cambridge: Cambridge UP, 1995.

Williams, George C. *Adaptation and Natural Selection: A Critique of Some Current Evolutionary Thought.* Princeton: Princeton UP, 1966.

Wilson, A. N. *God's Funeral.* New York: Norton, 1999.

Wilson, David Sloan. *Darwin's Cathedral: Evolution, Religion, and the Nature of Society*. Chicago: The University of Chicago, 2002.

Wilson, Edward O. *Consilience: The Unity of Knowledge*. New York: Knopf, 2003.

――― *Biophilia*. Cambridge, Mass.: Harvard UP, 1984.

――― *In Search of Nature*. Washington, D. C.: Island, 1996.

――― *On Human Nature*. Cambridge, Mass.: Harvard UP, 1978.

――― *Sociobiology: The New Synthesis*. 1975. 25th anniversary ed. Cambridge, Mass.: Belknap/Harvard UP, 2000.

Wright, Robert. *The Moral Animal: Evolutionary Psychology and Everyday Life*. New York: Pantheon, 1994.

Ziman, John. *Real Science: What It Is, and What It Means*. Cambridge: Cambridge UP, 2000.

Zimmer, Carl. *Evolution: The Triumph of an Idea*. New York: Harper Collins, 2001.

Index

Adaptation n77, 102-107, 131, 140, 158, 161f, 188, 197f
Anthropocentrism 61, 130, 141, 212ff
Aristotle 19, 44f, n54, 174, 180

Barth, Karl 41

Common descent 96
Complexity 1, 4f, 7, 32, 41, 44, 49, 62, 72f, 105, 113f, 118f, 121f, 130, 141, 143, 165, 167, 172f, 177ff, 187, n192, 206
Conradi, Peter 11
Culture 1, 5, 27, 35, 57, n60, 61f, 84, 94, 112, 142, 159, 162, 164, n169, n181, 211

Darwin, Charles 3, 6f, 10, 46f, 53f, 69, 71ff, esp. 89-158, 167-170, 171-216
Darwinism 6, 21, 26, 48, 89, 111, 136, 214ff
Dawkins, Richard 3, 19, 24f, n65, n97, 99, n109, 111, n114, 127, 142-147, n155, 159, n161, n190, n206
Dennett, Daniel 46, 91f, 94, n159, n161, n197
Descartes, René 44f, 53, 55f, n57, 184f
Dobzhansky, Theodosius 97, 108f

Emergence 118-123, 167
Epistemology 1-4, 6, 49, 80, 169, 171, 182, 193, 198, 200, 214
   Darwinian 174-189
   Evolutionary 1, 3, 169, 192f, 198ff, 214
   Process 81
Ethology 38, 139
Evolution, Darwinian 1ff, 5ff, 21, 37, 46, n77, esp. 89-128, 130f, 136, 142, 158f, 170, 180, 193, 196f, 208, 214
Ethics 4, 12-21, 48, n49, 50, 60, 73, 130, 213

Gaia theory 20, 63-69, 205
Genetics 24, 60, n65, 97f, n108, 112, n114, 123, 139

God 2, 21, 31f, n42, n49, n67, 132, 137ff, 143, 156-159, 166ff, n180, 183, 186, 200, 203-206, 209-213

Haught, John n77, n78, 153f, 167ff, 208
Hegel, G. W. F. 19, n44
Hobbes, Thomas 45
Humanism 52, 58f, 61, 141
Hume, David 45, n73, 172, 189

Individualism 36-39, 65, 190

James, William n67, 173

Kant, Immanuel 14

Mayr, Ernst n90, 94, n96, 97, 114, 117
Metaphysics 4, 12ff, 17, n46, 50, 55, 122, 130, 153f, 195
Midgley, Mary *passim*
Monod, Jacques 146, 152f
Moore, G. E. 12f, n15
Murdoch, Iris 11, 13-18, 33
McKinnon, Donald 13

Natural selection 32, 69, 91f, 96, 99-104, 107-110, 112ff, 125, 131, n135, 139, 143, 158ff, 188, 194f, 210f, 214f

Objectivism 153

Parsimony 119, 175, 191
Philosophy 5, 7, 9-19, 21f, 28, 30, 33, n40, 42, 44f, 48, 56, 58, 61, 71, 80, 82, 127, 129, 136, 148, 166, 170, n180, 185, 205, n209
  Empiricist 189
  Modern 15
  Moral 12, 14, 17
  Natural 71, 89, 120, 135, 175, 185
  Post-modern 16
Plato 44-45
  Platonic ideas 13

Rahner, Karl 182
Reductionism 21, 30, 32, n121
  also, Reductivism 187, 190f
Religion *passim*

Science *passim*
  Science and Religion 4, 49, 68, 70-87, 108, 127, 137, 174, 205f
Selfish gene(s) 65, 190, 196

Simplicity 30ff, 72, 81, 213
Sociobiology 3, 68, 139f, 145, 151
Spinoza 174, n180
Subjectivism 153

Theology n1, 41, 45, n49, 56, 77, 80, 82, 128f, 135, 169f, 180ff, 208

Variation n77, 92f, 95f, 100, n116, 117f, 124, 131, 134f, 188, 198

Warnock, Mary 11
Wilson, E. O. 3, 19, n27, 68f, 139-142, 144f, n147, 159

Ziman, John 163, 197ff

www.ingramcontent.com/pod-product-compliance
Lightning Source LLC
Chambersburg PA
CBHW070941230426
43666CB00011B/2513